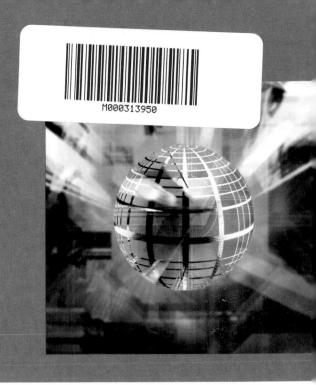

SIGNIFICANT CHANGES TO THE

INTERNATIONAL RESIDENTIAL CODE

2009 EDITION

DELMAR
CENGAGE Learning

Australia • Brazil • Japan • Korea • Mexico • Singapore • Spain • United Kingdom • United States

DELMAR
CENGAGE Learning™

Significant Changes to the International Residential Code 2009 Edition International Code Council

Delmar Staff:

Vice President, Technology and Trades Professional Business Unit: Gregory L. Clayton

Product Development Manager: Ed Francis

Development: Nobina Chakraborti

Director of Marketing: Beth A. Lutz

Executive Marketing Manager: Taryn Zlatin

Marketing Manager: Marissa Maiella

Production Director: Carolyn Miller

Production Manager: Andrew Crouth

Content Project Manager: Andrea Majot

Art Director: Benjamin Gleeksman

ICC Staff:

Senior Vice President, Business and Product Development: Mark A. Johnson

Technical Director, Product Development: Doug Thornburg

Manager, Project and Special Sales: Suzane Nunes Holten

Senior Marketing Specialist: Dianna Hallmark

For product information and technology assistance, contact us at **Professional Group Cengage Learning Customer & Sales Support, 1-800-354-9706**

For permission to use material from this text or product, submit all requests online at **www.cengage.com/permissions.** Further permissions questions can be e-mailed to **permissionrequest@cengage.com.**

Library of Congress Control Number: 2009921510

ISBN-13: 978-1-4354-0122-8

ISBN-10: 1-4354-0122-0

ICC World Headquarters
500 New Jersey Avenue, NW
6th Floor
Washington, D.C. 20001-2070
Telephone: 1-888-ICC-SAFE (422-7233)
Website: **http://www.iccsafe.org**

Delmar
5 Maxwell Drive
Clifton Park, NY 12065-2919
USA

Cengage Learning is a leading provider of customized learning solutions with office locations around the globe, including Singapore, the United Kingdom, Australia, Mexico, Brazil and Japan. Locate your local office at: **international.cengage.com/region**

Cengage Learning products are represented in Canada by Nelson Education, Ltd.

Visit us at **www.InformationDestination.com.**
For more learning solutions, **visit www.cengage.com**

Notice to the Reader

Publisher does not warrant or guarantee any of the products described herein or perform any independent analysis in connection with any of the product information contained herein. Publisher does not assume, and expressly disclaims, any obligation to obtain and include information other than that provided to it by the manufacturer. The reader is expressly warned to consider and adopt all safety precautions that might be indicated by the activities described herein and to avoid all potential hazards. By following the instructions contained herein, the reader willingly assumes all risks in connection with such instructions. The publisher makes no representations or warranties of any kind, including but not limited to, the warranties of fitness for particular purpose or merchantability, nor are any such representations implied with respect to the material set forth herein, and the publisher takes no responsibility with respect to such material. The publisher shall not be liable for any special, consequential, or exemplary damages resulting, in whole or part, from the readers' use of, or reliance upon, this material.

Printed in the United States
1 2 3 4 5 XX 11 10 09

Contents

Preface

The purpose of *Significant Changes to the International Residential Code, 2009 Edition*, is to familiarize building officials, fire officials, plans examiners, inspectors, design professionals, contractors, and others in the building construction industry with many of the important changes in the 2009 IRC. This publication is designed to assist those code users in identifying the specific code changes that have occurred and, more importantly, understanding the reasons behind the changes. It is also a valuable resource for jurisdictions in their code-adoption process.

Only a portion of the total number of code changes to the IRC are discussed in this book. The changes selected were identified for a number of reasons, including their frequency of application, special significance, or change in application. However, the importance of those changes not included is not to be diminished. Further information on all code changes can be found in the *Code Changes Resource Collection*, published by the ICC. The resource collection provides the published documentation for each successful code change contained in the 2009 IRC since the 2006 edition.

Significant Changes to the International Residential Code, 2009 Edition, is organized into nine parts, each representing a distinct grouping of code topics. It is arranged to follow the general layout of the IRC, including code sections and section number format. Throughout the book, each change is accompanied by either a photograph or an illustration to assist and enhance the reader's understanding of the specific change. A summary and a discussion of the significance of the changes are also provided. Each code change is identified by type, be it an addition, modification, clarification, or deletion.

The code change itself is presented in a format similar to the style utilized for code-change proposals. Deleted code language is shown with a strike-through, and new code text is indicated by underlining. As a result, the actual 2009 code language is provided, as well as a comparison with the 2006 language, so the user can easily determine changes to the specific code text.

As with any code-change text, *Significant Changes to the International Residential Code, 2009 Edition*, is best used as a study companion to the 2009 IRC. Because only a limited discussion of each change is provided, the code itself should always be referenced in order to gain a more comprehensive understanding of the code change and its application.

The commentary and opinions set forth in this text are those of the author and do not necessarily represent the official position of the ICC. In addition, they may not represent the views of any enforcing agency, as such agencies have the sole authority to render interpretations of the IRC. In many cases, the explanatory material is derived from the reasoning expressed by the code-change proponent.

Comments concerning this publication are encouraged and may be directed to the ICC at *significantchanges@iccsafe.org*.

About the International Residential Code

Building officials, design professionals, contractors, and others involved in the field of residential building construction recognize the need for a modern, up-to-date residential code addressing the design and installation of building systems through requirements emphasizing performance. The *International Residential Code*® (IRC), in the 2009 edition, is intended to meet these needs through model code regulations that safeguard the public health and safety in all communities, large and small. The IRC is kept up to date through the open code-development process of the International Code Council (ICC). The provisions of the 2006 edition, along with those code changes approved through 2008, make up the 2009 edition.

The ICC, publisher of the IRC, was established in 1994 as a nonprofit organization dedicated to developing, maintaining, and supporting a single set of comprehensive and coordinated national model construction codes. Its mission is to provide the highest quality codes, standards, products, and services for all concerned with the safety and performance of the built environment.

The IRC is one in a family of International Codes® published by the ICC. This comprehensive residential code establishes minimum regulations for residential building systems by means of prescriptive and performance-related provisions. It is founded on broad-based principles that make possible the use of new materials and new building designs. The IRC is a comprehensive code containing provisions for building, energy conservation, mechanical, fuel gas, plumbing, and electrical systems. The IRC is available for adoption and use by jurisdictions internationally. Its use within a governmental jurisdiction is intended to be accomplished through adoption by reference, in accordance with proceedings establishing the jurisdiction's laws.

Acknowledgments

Grateful appreciation is due many ICC staff members for their generous assistance in the preparation of this publication: John Henry, PE, for his expert review and editing related to the structural provisions; Peter Kulczyk for his thorough review, corrections, and comments on the text and for sharing his photographs; Scott Stookey for welcome expertise, commentary, and photographs related to fire resistance and fire protection systems; Jay Woodward for providing timely information and feedback; Hamid Naderi, PE and Doug Thornburg, AIA for their usual expert direction and advice. Thanks also go to Don Sivigny, Construction Codes and Licensing Division, State of Minnesota, for his assistance with the energy efficiency provisions. All contributed to the accuracy and quality of the finished product.

About the Author

Stephen A. Van Note, CBO

International Code Council

Senior Technical Staff, Product Development

Stephen A. Van Note is a member of the senior technical staff of the International Code Council (ICC), where, as part of the Product Development team, he is responsible for authoring technical resource materials in support of the International Codes. His role also includes the management, review and technical editing of publications authored by outside sources. Prior to joining ICC in 2006, Van Note was building official for Linn County, Iowa. He has 15 years experience in code administration and enforcement, and more than 20 years experience in the construction field including project planning and management for residential, commercial, and industrial buildings. A certified building official and plans examiner, Van Note also holds certifications in five inspection categories.

SIGNIFICANT CHANGES TO THE

INTERNATIONAL RESIDENTIAL CODE

2009 EDITION

PART 1

Scope and Administration

■ **Chapter 1** Scope and Administration

The administration part of the *International Residential Code (IRC)* covers the general scope, purpose, applicability, and other administrative issues related to the regulation of residential buildings by building safety departments. The administrative provisions are the guiding light for the entire code and make clear the responsibilities and duties of various parties involved in residential construction and the applicability of technical provisions within a legal, regulatory, and code enforcement arena.

Section R101.2 establishes the criteria for buildings that are regulated under the IRC. Buildings beyond the scope of Section R101.2 are regulated by the *International Building Code (IBC)*. The IRC governs detached one- and two-family dwellings and townhouses that are not more than three stories above grade plane in height and have their own separate means of egress. Buildings accessory to such buildings are also regulated by the IRC. A new exception permits live/work units in buildings constructed under the IRC.

The remaining topics in Part 1 deal with subjects such as work exempt from permits, and submittal and approval of construction documents. ■

R101.2
Scope, Grade Plane

R101.2
Scope, Live/Work Units

R105.2
Work Exempt from Permit

R106.1.1
Information on Construction Documents

R106.3.1
Approval of Construction Documents

R101.2

Scope, Grade Plane

CHANGE TYPE: Modification

CHANGE SUMMARY: *Grade plane* replaces the word *grade* in determining the story limitations of the IRC.

2009 CODE: R101.2 Scope. The provisions of the *International Residential Code for One- and Two-family Dwellings* shall apply to the construction, alteration, movement, enlargement, replacement, repair, equipment, use and occupancy, location, removal and demolition of detached one- and two-family dwellings and townhouses not more than three stories above grade <u>plane</u> in height with a separate means of egress and their accessory structures.

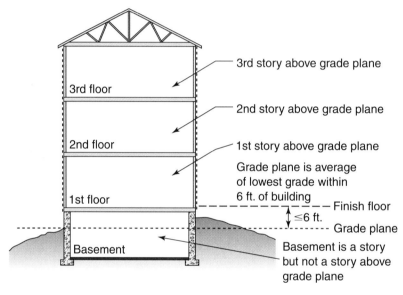

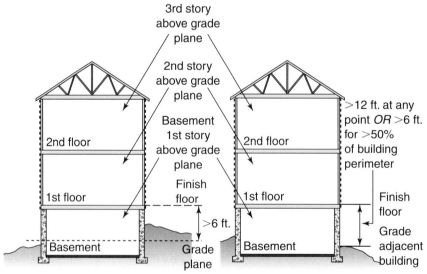

3 stories above grade plane

R202 Definitions
Story above Grade <u>Plane</u>.   Any story having its finished floor surface entirely above grade <u>plane</u>, except that a basement shall be considered as a story above grade <u>plane</u> where the finished surface of the floor above the basement meets any one of the following:

1. Is more than 6 feet (1829 mm) above grade plane.
2. Is more than 6 feet (1829 mm) above the finished ground level for more than 50 percent of the total building perimeter.
3. Is more than 12 feet (3658 mm) above the finished ground level at any point.

CHANGE SIGNIFICANCE: Grade and grade plane are defined terms with different meanings and application. Grade is simply the finished ground level adjoining the building at exterior walls. Grade plane on the other hand is an average of the finished ground level measured at the lowest point within six feet of the exterior wall (unless the property line is closer than six feet).

Previously, the scope of the IRC limited dwellings to three stories above grade. This generally meant three stories entirely above the adjoining ground level at any point around the perimeter of the building. The exception to the general rule was a partially below grade basement considered by definition to be a story above grade when it met one of three criteria related to grade plane or finished ground level. The change to grade plane intends to clarify the application and to bring consistency in terminology both within the IRC and in relation to the scope of the *International Building Code* (IBC). For detached one- and two-family dwellings and townhouses, along with their accessory structures, IBC Section 101.2 sends the code user to the IRC for such buildings not more than three stories above grade plane in height. The new language in the IRC will now match the scoping provisions of the IBC.

In practice, the new language will not change the outcome in most cases. Typically, basements considered a story above *grade* in the 2006 IRC will also be a story above *grade plane* in the 2009 IRC because the three criteria in the revised definition have not changed. When the basement is not a story above grade plane, the code still permits three stories in addition to the basement.

R101.2

Scope, Live/Work Units

CHANGE TYPE: Addition

CHANGE SUMMARY: A new exception to the scope of the IRC references the *International Building Code* (IBC) for provisions on *live/work units*, a mix of residential and non-residential uses. The intent of this change is to permit live/work units in one- and two-family dwellings and townhouses constructed under the IRC, provided such units comply with the specific requirements in Section 419 of the IBC.

2009 CODE: R101.2 Scope.   The provisions of the *International Residential Code for One- and Two-family Dwellings* shall apply to the construction, alteration, movement, enlargement, replacement, repair, equipment, use and occupancy, location, removal and demolition of detached one- and two-family dwellings and townhouses not more than three stories above grade <u>plane</u> in height with a separate means of egress and their accessory structures.

> **Exception:** <u>Live/work units complying with the requirements of Section 419 of the *International Building Code* shall be permitted to be built as one- and two-family dwellings or townhouses. Fire suppression required by Section 419.5 of the *International Building Code* when constructed under the *International Residential Code for One- and Two-family Dwellings* shall conform to Section 903.3.1.3 of the *International Building Code.*</u>

CHANGE SIGNIFICANCE: A *live/work unit* is defined in the IBC as a dwelling unit or sleeping unit in which a significant portion of the space includes a nonresidential use that is operated by the tenant. The added language in the IRC recognizes live/work units constructed as one- and two-family dwellings and townhouses, but sends the user to IBC Section 419 for the specific limitations and requirements that apply to both the dwelling portion and the work portion of the unit. As such, live/work units are limited to the following:

• A maximum combined area of 3,000 square feet including the residential space and the work space in the live/work unit

- A nonresidential area not greater than 50 percent of the live/work unit area
- Nonresidential functions on the main floor only
- At any one time, a maximum of five workers who do not reside in the dwelling
- Storage area not greater than 10 percent of the nonresidential space

The IBC provisions for structural loading, means of egress, accessibility for persons with disabilities, fire alarm systems, and fire sprinkler systems apply to live/work units. For live/work units in one- and two-family dwellings and townhouses constructed under the IRC, an NFPA 13-D sprinkler system is permitted to satisfy the fire suppression requirements.

The live/work unit concept permits both residential and commercial functions within a dwelling unit without fire-resistance-rated separations between the uses or stair enclosures. This concept of design and construction allows a public service business with employees working within a residence and allows the public to enter the work area of the unit to acquire service. Beauty salons, coffee shops, and chiropractor's offices are examples of commercial establishments in a live/work unit. A home office that comprises no more than 10 percent of the dwelling unit does not create a live/work unit under the IBC definition. For example, a small home office for an architect or consultant would not be subject to the requirements of Section 419 of the IBC.

R105.2

Work Exempt from Permit

CHANGE TYPE: Modification

CHANGE SUMMARY: The floor area for accessory structures that are exempt from permits has increased from 120 to 200 square feet. Decks that are not attached to the structure and are limited in area and height are now included in the specific list of permit exemptions. The code also now provides a list of electrical work that qualifies for exemption from the permit requirements.

2009 CODE: R105.2 Work Exempt From Permit. Permits shall not be required for the following. Exemption from permit requirements of this code shall not be deemed to grant authorization for any

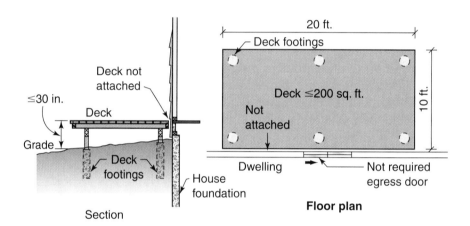

Deck exempt from permit

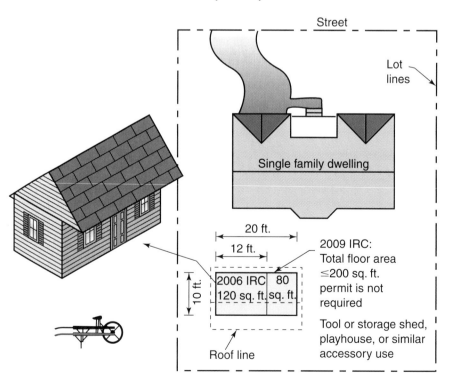

Sheds exempt from permits

work to be done in any manner in violation of the provisions of this code or any other laws or ordinances of this jurisdiction.

Building:

1. One-story detached accessory structures used as tool and storage sheds, playhouses and similar uses, provided the floor area does not exceed ~~120~~ 200 square feet (~~11.15~~ 18.58 m^2).

2. through **9.** (No change to text.)

10. Decks not exceeding 200 square feet (18.58 m^2) in area, that are not more than 30 inches (762 mm) above grade at any point, are not attached to a dwelling, and do not serve the exit door required by Section R311.4.

Electrical:

1. Listed cord and plug connected temporary decorative lighting.

2. Reinstallation of attachment plug receptacles but not the outlets therefor.

3. Replacement of branch circuit overcurrent devices of the required capacity in the same location.

4. Electrical wiring, devices, appliances, apparatus or equipment operating at less than 25 volts and not capable of supplying more than 50 watts of energy.

5. **Repairs and maintenance: A permit shall not be required for** Minor repair work, including the replacement of lamps or the connection of approved portable electrical equipment to approved permanently installed receptacles.

CHANGE SIGNIFICANCE: The maximum size of one-story detached accessory structures not requiring a permit has increased to 200 square feet of floor area, a limitation that matches previous provisions of the 2000 and 2003 IRC. The exemption was reduced to a floor area of 120 square feet in the 2006 code for consistency with the IBC provisions and in response to a concern that 200 square feet was large enough to accommodate a garage or other use that would increase the hazard and justify closer scrutiny. Proponents for reverting to the 200 square foot allowance, a common size for manufactured sheds, cited a lack of technical justification for the reduced size and a good track record for the larger sheds. In addition, a larger size than that allowed by the IBC was deemed appropriate for IRC-regulated buildings posing less hazard due to the nature of the use and the typical materials in storage.

Small decks join the list of items that are exempt from the permitting requirements. To qualify for the exemption, the deck must meet all four of the following criteria:

- Not exceed 200 square feet in area
- Not be more than 30 inches above grade at any point
- Not be attached to a dwelling
- Not serve the required exit door

R105.2 continues

R105.2 continued   Similar to sheds that do not exceed 200 square feet in area, small decks that are close to the ground do not pose a significant risk to occupant safety. In addition, self-supporting decks eliminate the concerns of improper attachment to the structure. While the items listed in Section R105.2 are exempt from permits, construction must still comply with the minimum requirements of the code.

The previous language for the electrical work exemption from permits, basically any repair work of a minor nature, was judged to be vague and subjective. The code now provides a list of repairs and installations considered sufficiently routine to forgo the permitting and inspection process. Minor repair work remains in the list, giving discretion to the building official to make a determination on work that is not otherwise specifically mentioned.

R106.1.1

Information on Construction Documents

CHANGE TYPE: Modification

CHANGE SUMMARY: The code now lists specific wall bracing information to be included on drawings or other construction documents. As with other submittals, the building official is authorized to decide if such information is necessary for a particular project.

2009 CODE: R106.1.1 Information on Construction Documents. Construction documents shall be drawn upon suitable material. Electronic media documents are permitted to be submitted when approved by the building official. Construction documents shall be of sufficient clarity to indicate the location, nature, and extent of the work proposed and show in detail that it will conform to the provisions of this code and relevant laws, ordinances, rules and regulations, as determined by the building official. <u>Where required by the building official, all braced wall lines shall be identified on the construction documents, and all pertinent information including but not limited to bracing methods, location, and length of braced wall panels, foundation requirements, and attachment of braced wall panels at top and bottom shall be provided.</u>

CHANGE SIGNIFICANCE: With increasing complexity of the design and construction of residential buildings, and the increasing number of options and flexibility in the prescriptive wall bracing provisions, it is often difficult for the plan reviewer to determine compliance in the absence of such bracing information in the submittal documents. Requiring wall bracing details on construction drawings ensures that bracing is being considered during the design and review process. It

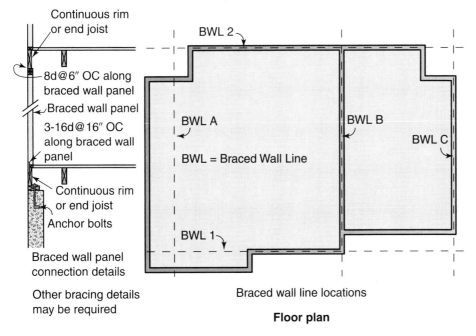

Braced wall panel connection details

Other bracing details may be required

Braced wall line locations

Floor plan

Where required by the building official, wall bracing information must be shown on construction documents.

R106.1.1 continues

R106.1.1 continued will also make it easier for the builder to properly construct the required bracing on the job site when the details are clearly spelled out on the drawings. The new provisions require the location of all braced wall lines, and the type, location, and length of the braced wall panels to be indicated on the plans. In addition, foundation and attachment details related to wall bracing must be identified.

For simple rectangular shapes, the location of the braced wall lines may be obvious, and the additional bracing information is not required to make a determination. However, when houses have irregular shapes, the answers are not so obvious, and multiple solutions could be conceived. In this case the building official is specifically authorized to require such information on the submittal documents. This is a shift of responsibility from the plan reviewers to the plan preparers in order to eliminate doubt and to ensure adequate wall bracing to meet the minimum requirements of the code.

CHANGE TYPE: Modification

CHANGE SUMMARY: The language for written approval on construction documents now matches the corresponding section of the *International Building Code* (IBC).

2009 CODE: R106.3.1 Approval of Construction Documents. When the building official issues a permit, the construction documents shall be approved in writing or by a stamp which states "~~APPROVED PLANS PER IRC SECTION R106.3.1~~ REVIEWED FOR CODE COMPLIANCE." One set of construction documents so reviewed shall be retained by the building official. The other set shall be returned to the applicant, shall be kept at the site of work and shall be open to inspection by the building official or his or her authorized representative.

CHANGE SIGNIFICANCE: Prior to the 2006 IRC, Section R106.3.1 required approval of construction drawings in writing or by stamp without stating any specific wording. The 2006 IRC introduced specific approval language that included the applicable code section. Recognizing that building departments and plans examiners review plans for buildings regulated by the IRC, IBC, and other International Codes, the IRC now requires the approved plans to indicate that they have been reviewed for code compliance. The new generic language promotes consistency between the IBC and IRC and within the particular building department.

R106.3.1

Approval of Construction Documents

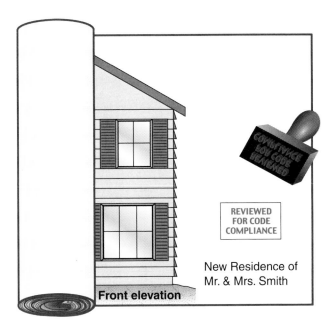

Approval of construction documents

PART 2
Definitions

■ **Chapter 2** Definitions

The definitions contained within the IRC are intended to reflect the special meaning of such terms within the scope of the code. As terms can often have multiple meanings within their ordinary day-to-day use or within the various disciplines of the construction industry, it is imperative that their meaning within the context of the IRC be given.

Section R201.3 requires that where terms are not defined in the IRC, the meaning for such a term be taken from the other codes published by the International Code Council (ICC), such as the IBC or others in the family of codes developed by the ICC (I-Codes). Section R201.4 requires that a term not defined anywhere within the family of I-Codes must have the ordinarily accepted meaning such as that which the context implies.

The IRC definitions are contained within Chapter 2; however, terms specifically related to fuel gas and electrical systems are defined in specific chapters, Chapter 24 for fuel gas and Chapter 35 for electrical. Terms defined in Chapters 24 and 35 are not necessarily repeated within Chapter 2. ■

R202

Definitions, Attic and Habitable Attic

R202

Definitions, Labeled and Listed

R202

Definitions, Structural Insulated Panel (SIP)

R202

Definitions, Attic and Habitable Attic

CHANGE TYPE: Addition

CHANGE SUMMARY: The definition for attic has been revised and a new definition for habitable attic has been added.

2009 CODE:

Attic. The unfinished space between the ceiling ~~joists~~ <u>assembly</u> of the top story and the roof ~~rafters~~ <u>assembly</u>.

Attic, Habitable. <u>A finished or unfinished area, not considered a story, complying with all of the following requirements:</u>

1. <u>The occupiable floor area is at least 70 square feet (6.5 m^2), in accordance with Section R304,</u>
2. <u>The occupiable floor area has a ceiling height in accordance with Section R305,</u>
3. <u>The occupiable space is enclosed by the roof assembly above, knee walls (if applicable) on the sides, and the floor-ceiling assembly below.</u>

CHANGE SIGNIFICANCE: The revised definition for attic is more inclusive in recognizing that elements other than ceiling joists and rafters may enclose the attic space. For example, manufactured storage or room trusses are often used for roof-ceiling construction. In this case, the attic space is between the top and bottom chords of the trusses.

A new defined term in the 2009 IRC, a *habitable attic* is occupiable space between the floor/ceiling assembly and the roof assembly. The major significance of this change is that habitable attics are not considered a story. The code now effectively permits five habitable levels for one- and two-family dwellings and townhouses—a basement below grade plane, three stories above grade plane, and a habitable attic. Each level is unlimited in size. Previously, an attic converted to habitable space would no longer be an attic and would be considered a story. Similarly in new construction under the 2006

R202 continues

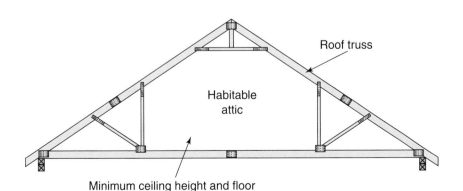

Habitable attic

Roof truss

Minimum ceiling height and floor area as required for habitable spaces.

Habitable attic

R202 continued   IRC, finished upper levels with vaulted ceilings, sometimes called bonus rooms, would qualify as stories. With the new definition, such areas are no longer considered stories.

Habitable attics must meet the minimum room size and ceiling height requirements for habitable spaces, and require a smoke detector (Section R314), emergency escape and rescue opening (Section R310), and a means of egress complying with Section R311.

CHANGE TYPE: Modification

CHANGE SUMMARY: Definitions for labeled and listed have been revised for clarity and consistency with the IBC and other International Codes.

2009 CODE: Labeled. ~~Devices, equipment or~~ Equipment, materials or products to which have been affixed a label, seal, symbol or other identifying mark of a nationally recognized testing laboratory, inspection agency or other organization concerned with product evaluation that maintains periodic inspection of the production of the above labeled items ~~that attests to compliance with a specific standard~~ and whose labeling indicates either that the equipment, material or product meets identified standards or has been tested and found suitable for a specified purpose.

Listed ~~and Listing~~. ~~Terms referring to equipment that is shown in a list published by an approved testing agency qualified and equipped for experimental testing and maintaining an adequate periodic inspection of current productions and whose listing states that the equip-~~

R202 continues

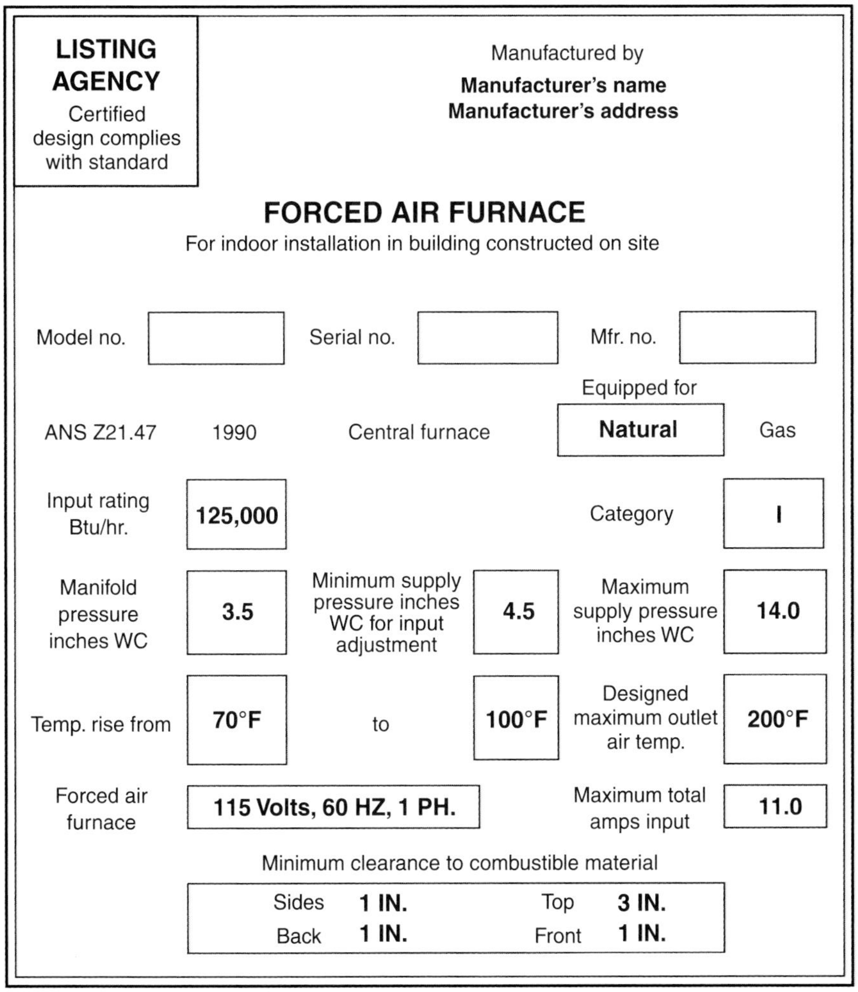

R202

Definitions, Labeled and Listed

Example of a label for a listed appliance

R202 continued   ~~ment complies with nationally recognized standards when installed in accordance with the manufacturer's installation instructions.~~

<u>Equipment, materials, products, or services included in a list published by an organization acceptable to the code official and concerned with evaluation of products or services that maintains periodic inspection of production of listed equipment or materials or periodic evaluation of services and whose listing states either that the equipment, material, product or service meets identified standards or has been tested and found suitable for a specified purpose.</u>

CHANGE SIGNIFICANCE: *Listed* and *labeled* are important terms used throughout the IRC to require that equipment, products, and materials perform the intended function in a safe manner. The building official relies on labeling and listing to verify code compliance. The revised definitions are now consistent with the IBC and others in the family of International Codes. The code now clarifies that the testing laboratory or testing agency must be nationally recognized. It also recognizes that not all products are tested to a specific standard but may prove through testing to be satisfactory for the intended function for which no standard applies.

R202

Definitions, Structural Insulated Panel (SIP)

CHANGE TYPE: Addition

CHANGE SUMMARY: The IRC now includes prescriptive methods of construction using structural insulated panels (SIPs). Section R202 has added a number of definitions relevant to SIP construction.

2009 CODE: Structural Insulated Panel (SIP). A structural sandwich panel that consists of a light-weight foam plastic core securely laminated between two thin, rigid wood structural panel facings.

Cap Plate. The top plate of the double top plates used in structural insulated panel (SIP) construction. The cap plate is cut to match the panel thickness such that it overlaps the wood structural panel facing on both sides.

Core. The light-weight middle section of the structural insulated panel composed of foam plastic insulation, which provides the link between the two facing shells.

R202 continues

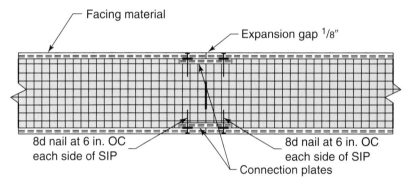

Surface spline connection

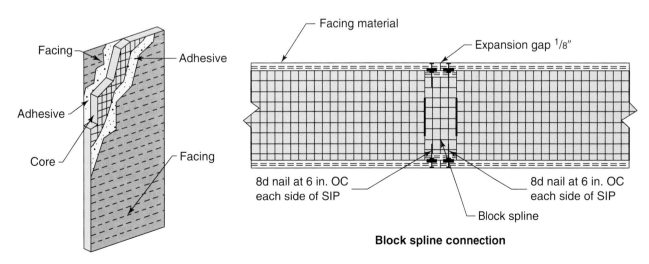

Structural insulated panel (SIP)

Spline connection for structural insulated panels (SIPs)

R202 continued

Facing.   The wood structural panel facings that form the two out-most rigid layers of the structural insulated panel.

Panel Thickness.   Thickness of core plus two layers structural wood panel facings.

Spline.   A strip of wood structural panel cut from the same material used for the panel facings, used to connect two structural insulated panels. The strip (spline) fits into a groove cut into the vertical edges of the two structural insulated panels to be joined. Splines are used behind each facing of the structural insulated panels being connected as shown in Figure R613.8.

CHANGE SIGNIFICANCE: Manufactured under factory controlled conditions, SIPs are typically made by sandwiching a core of foam plastic insulation between two facings of oriented strand board (OSB). Other rigid wood structural panel materials may also be used for the outside skin of the panel. SIP construction is an alternative to the conventional light-frame provisions for walls, floors, and roofs of residential buildings. Edges of the panels are joined with splines and sealed to provide a tight, energy-efficient building envelope.

PART 3

Building Planning and Construction

Chapters 3 through 10

Part 3 deals with the overall issues of building planning, design, site location, fire safety, egress, structural system, and other such related issues. Chapter 3 includes the bulk of nonstructural issues such as the location on the lot, fire-resistive construction, light and ventilation, emergency escape and rescue, fire protection, and many other issues. Section R308 covers glazing materials and the locations where safety glazing is required, and Sections R311 through R315 address means of egress, guards, automatic fire sprinkler systems, smoke alarms, and carbon monoxide alarms. In addition to such issues, Chapter 3 provides the overall structural loads for the design of residential buildings. Section R301, Design Criteria, addresses all structural loading issues of live loads, dead loads, and other environmental loads such as wind, seismic, and snow.

Other chapters within Part 3 address the prescriptive as well as the performance criteria for building foundations, floor construction, wall construction, wall coverings, roof construction, roof assemblies, chimneys, and fireplaces. ∎

R301.1.1
Alternative Provisions

R301.2.1.1
Design Criteria

R301.2.1.2 AND TABLE R301.2.1.2
Protection of Openings

R301.2.1.5 AND TABLE R301.2(1)
Topographic Wind Effects

R301.2.2
Seismic Provisions

R301.2.3
Snow Loads

R301.3
Story Height

TABLE R301.5
Minimum Uniformly Distributed Live Loads

R302.1 AND TABLE R302.1
Fire-resistant Construction at Exterior Walls

R302.2 AND R302.3
Dwelling Unit Separation

R302.4
Rated Penetrations for Dwelling Unit Separation

R302.5
Garage Openings and Penetrations

continues

20

R301.1.1

Alternative Provisions

CHANGE TYPE: Modification

CHANGE SUMMARY: The IRC now recognizes a recently developed standard for log construction. The referenced standard for cold-formed steel framing has been updated.

2009 CODE: R301.1.1 Alternative Provisions. As an alternative to the requirements in Section R301.1 the following standards are permitted subject to the limitations of this code and the limitations therein. Where engineered design is used in conjunction with these standards, the design shall comply with the *International Building Code.*

1. American Forest and Paper Association (AF&PA) *Wood Frame Construction Manual* (WFCM).
2. American Iron and Steel Institute (AISI) *Standard for Cold-Formed Steel Framing—Prescriptive Method for One- and Two-Family Dwellings* ~~(COFS/PM)~~ *~~with Supplement to Standard for Cold-Formed Steel Framing- Prescriptive Method for One- and Two-Family Dwellings~~* (AISI S230).
3. ICC-400 Standard on the Design and Construction of Log Structures.

CHANGE SIGNIFICANCE: Log home construction is not uncommon in some regions of the country. The 2006 IRC introduced stress grading and certification requirements for log structural members for floor, wall, and roof framing in Sections R502.1.6, R602.1.3, and R802.1.5. The 2009 IRC references a standard that was developed to be consistent with the scope of the ICC family of codes in adequately protecting the public health, safety, and welfare. The new ICC-400 standard re-

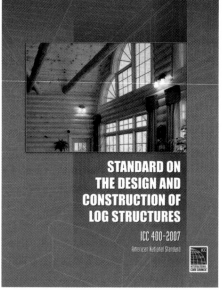

STANDARD ON THE DESIGN AND CONSTRUCTION OF LOG STRUCTURES
ICC 400-2007
American National Standard

flects industry consensus and provides nationally recognized requirements for the design and construction of log structures. The standard includes provisions for log grading, moisture content, structural design, settling allowances, fire resistance, and thermal envelope properties of log structures.

The 2007 edition of AISI S230 replaces the 2004 edition of AISI-PM and includes some notable changes. The prescriptive methods for cold-formed steel framing now apply to three-story buildings, an increase from an allowable two stories in the previous standard and the 2006 IRC, and are consistent with the height limits of conventional wood frame construction. The new standard defines maximum story height and clarifies the provisions for irregular buildings located in a high seismic or high wind area. Wall stud, rafter, and header tables have been updated to reflect current data. The standard also adds or revises structural provisions related to anchor bolt washers, bearing stiffeners, connections, braced walls, and design of gable end walls.

R301.2.1.1
Design Criteria

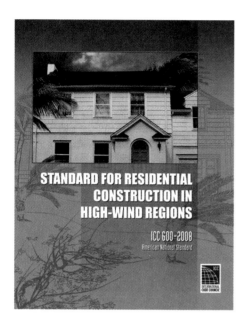

CHANGE TYPE: Modification

CHANGE SUMMARY: The IRC now recognizes structural insulated panel (SIP) construction for high wind areas, bringing the list of design alternatives to six. The new ICC-600 standard for high wind areas has replaced the legacy code standard SSTD 10. Additional text clarifies that building elements not addressed in the referenced methods of construction need to comply with the provisions of the IRC.

2009 CODE: R301.2.1.1 Design Criteria. ~~Construction~~ In regions where the basic wind speeds from Figure R301.2(4) equal or exceed 100 miles per hour (45 m/s) in hurricane-prone regions, or 110 miles per hour (49 m/s) elsewhere, <u>the design of buildings</u> shall be ~~designed~~ in accordance with one of the following <u>methods</u>. <u>The elements of design not addressed by those documents in Items 1 through 4 shall be in accordance with this code.</u>

1. American Forest and Paper Association (AF&PA) *Wood Frame Construction Manual for One- and Two-Family Dwellings* (WFCM); or
2. ~~*Southern Building Code Congress International Standard for Hurricane Resistant Residential Construction* (SSTD 10)~~; <u>*International Code Council (ICC) Standard for Residential Construction in High Wind Regions (ICC-600)*</u>; or
3. *Minimum Design Loads for Buildings and Other Structures* (ASCE 7); or
4. American Iron and Steel Institute (AISI), *Standard for Cold-Formed Steel Framing—Prescriptive Method for One- and Two-Family Dwellings* ~~*(COFS/PM) with Supplement to Standard for Cold-Formed Steel Framing—Prescriptive Method For One- and Two-Family Dwellings*~~ <u>(AISI S230)</u>.
5. Concrete construction shall be designed in accordance with the provisions of this code.
6. <u>Structural insulated panel (SIP) walls shall be designed in accordance with the provisions of this code.</u>

CHANGE SIGNIFICANCE: Prior to the 2006 edition, the IRC generally limited the prescriptive design of structures to those areas having a wind speed less than 110 mph. For wind speeds of 110 mph or greater, the code stipulated a design in accordance with one of the referenced standards, including an engineered design in accordance with ASCE 7, as referenced by the IBC. The 2006 IRC further lowered prescriptive wind design to less than 100 mph in hurricane prone regions. This change establishing different thresholds depending on the region was in response to observation and structural analyses of widespread hurricane damage compared with the effects of relatively localized and short-lived thunderstorm activity. While maintaining the wind speed criteria, the 2009 IRC clarifies the intent that the wind speed limits apply to only the prescriptive wind design for elements covered in the referenced standards. For other elements of the building, the provisions of the IRC apply.

The new ICC-600 *Standard for Residential Construction in High Wind Regions* replaces the legacy code standard SSTD 10. The older standard was based on fastest mile wind speeds and wind loads of the *Standard Building Code.* ICC-600 provides contemporary requirements, including design for three-second gust wind speeds that are consistent with the wind provisions of the IBC and ASCE 7 and that supplement the IRC provisions.

The 2004 edition of AISI-PM has been updated to the 2007 AISI S230 *Standard for Cold-Formed Steel Framing—Prescriptive Method for One- and Two-Family Dwellings.* Notably, the allowable number of stories using the prescriptive methods for cold-formed steel framing has increased from two- to three-story buildings. The new standard also revises structural provisions related to anchor bolt washers, bearing stiffeners, connections, braced walls, and design of gable end walls.

Previously, when the building location exceeded the threshold wind speeds, the IRC referenced four standards for high wind design, including an engineered design in accordance with ASCE 7 and compatible with design requirements in the IBC. It also recognized prescriptive concrete construction designed in accordance with other provisions of the IRC because such construction is considered adequate in resisting high wind events. The IRC now provides an additional option for structural insulated panel (SIP) wall construction. In accordance with Section R613.2, the prescriptive provisions for SIP wall construction are limited to sites with a maximum design wind speed of 130 mph in Exposure A, B, or C and are not subject to the design wind speed limitations of Section R302.1.1.

R301.2.1.2 and Table R301.2.1.2

Protection of Openings

CHANGE TYPE: Modification

CHANGE SUMMARY: Protection of glazed openings for garage doors is now specifically required in windborne debris regions. When wood structural panels are used for any opening protection, they must be predrilled, and the mounting hardware must be permanently attached to the building to ease installation. The prescriptive methods for attaching wood structural panels now require additional anchors with greater embedment depth and resistance to pullout to significantly increase the safety factor.

2009 CODE: **R301.2.1.2 Protection of Openings.** Windows in buildings located in windborne debris regions shall have glazed openings protected from windborne debris. Glazed opening protection for windborne debris shall meet the requirements of the Large Missile Test of an approved impact resisting standard or ASTM E 1996 and ASTM E 1886 referenced therein. <u>Garage door glazed opening protection for windborne debris shall meet the requirements of an approved impact resisting standard or ANSI/DASMA 115.</u>

Exception: Wood structural panels with a minimum <u>thickness</u> of 7/16 inch (11 mm) and a maximum span of 8 feet (2438 mm) shall be permitted for opening protection in one- and two-story buildings. Panels shall be precut ~~so that they shall be~~ <u>and</u> attached to the framing surrounding the opening containing the product with the glazed opening. Panels <u>shall be predrilled as required for the anchorage method and</u> shall be secured with the attachment hardware provided. Attachments shall be designed

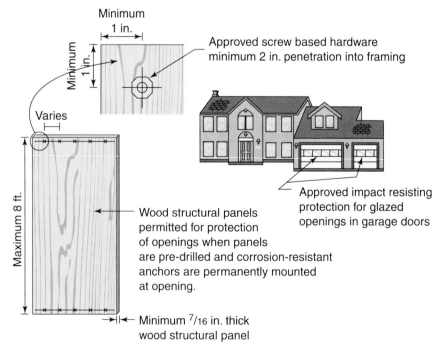

Protection of openings in windborne debris regions

TABLE R301.2.1.2 **Windborne Debris Protection Fastening Schedule for Wood Structural Panels**[a,b,c,d]

FASTENER TYPE	FASTENER SPACING (inches)[a,b]		
	Panel span ≤ 4 foot	4 feet < panel span ≤ 6 feet	6 feet < panel span ≤ 8 feet
No. ~~6~~ 8 wood ~~screws~~ based anchor with 2-inch embedment length	16	~~12~~ 10	~~9~~ 8
No. ~~8~~ 10 wood ~~screws~~ based anchor with 2-inch embedment length	16	~~16~~ 12	~~12~~ 9
¼-inch lag screw based anchor with 2-inch embedment length	16	16	16

For SI: 1 inch = 25.4 mm, 1 foot = 304.8 mm, 1 pound = 4.448N, 1 mile per hour = 0.447 m/s.

a. This table is based on 130 mph wind speeds and a 33-foot mean roof height.

b. Fasteners shall be installed at opposing ends of the wood structural panel. Fasteners shall be located a minimum of 1 inch from the edge of the panel.

c. ~~Anchors~~ Fasteners shall ~~be long enough to~~ penetrate through the exterior wall covering ~~and a minimum of 1¼ inches into wood wall framing and a minimum of 1¼ inches into concrete block or concrete, and into steel framing a minimum of 3 exposed threads~~ with an embedment length of 2 inches minimum into the building frame. Fasteners shall be located a minimum of 2½ inches from the edge of concrete block or concrete.

d. Where ~~screws~~ panels are attached to masonry or masonry/stucco, they shall be attached using vibration-resistant anchors having a minimum ultimate withdrawal capacity of ~~490~~ 1500 pounds.

R301.2.1.2 and Table R301.2.1.2 continues

to resist the component and cladding loads determined in accordance with either Table R301.2(2) or ~~Section 1609.6.5 of the International Building Code~~ ASCE 7, with the permanent corrosion-resistant attachment hardware provided and anchors permanently installed on the building. Attachment in accordance with Table R301.2.1.2 is permitted for buildings with a mean roof height of 33 feet (10 058 mm) or less where wind speeds do not exceed 130 miles per hour (58 m/s).

CHANGE SIGNIFICANCE: Previously, the IRC required glazing only in windows to be protected from windborne debris and did not specifically address protection of glazing in garage doors. The revision clarifies that such protection for impact resistance is required for glazing in garage doors in addition to windows. ANSI/DASMA 115 *Standard Method of Testing Garage Doors: Determination of Structural Performance Under Missile Impact and Cyclic Wind Pressure* is published by the Door & Access Systems Manufacturers Association specifically for the windborne debris resistance testing of garage doors and is considered the industry standard for this application. In addition to glazing protection, testing of the garage door in accordance with ANSI/DASMA 115 is also used to determine compliance with the component and cladding loads of Table R301.2(2). Section R301.2.1.2 also recognizes other approved standards demonstrating adequate impact resistance for garage door glazing.

There are many options for protecting glazed openings in windborne debris regions. A window fitted with approved impact-resistant

R301.2.1.2 and Table R301.2.1.2 continued

glass, for example, satisfies the requirement if tested to the prescribed standards. Manufactured shutters, panels, screens, and other devices meeting the large missile test are also an option. The exception to Section R301.2.1.2 allows a prescriptive method without testing that utilizes wood structural panels for glazed opening protection. The first part of the change to the exception addresses the concern of the difficulty of locating anchors and installing these precut panels during an advancing storm. The code now requires the panels to be predrilled for the attachment anchors, with the corrosion-resistant hardware provided and anchors permanently installed on the building. The intent is to require permanently mounted hardware so the panels can be quickly and easily installed.

The second part of this change significantly increases the wind resistance capacity of the panel attachment to the building. Larger diameter anchors are required, and the minimum embedment depth has increased from 1¼ inches to 2 inches. Anchor spacing for panels exceeding a 4-foot span has decreased, thereby increasing the number of anchors required. In addition, the minimum required capacity of masonry anchors has increased from 490 pounds to 1500 pounds to increase the safety factor.

CHANGE TYPE: Addition

CHANGE SUMMARY: Under very limited circumstances in localized geographic areas, design of buildings sited on a hill, ridge, or escarpment must consider the effects of topographic wind speedup.

2009 CODE: R301.2.1.5 Topographic Wind Effects. In areas designated in Table R301.2(1) as having local historical data documenting structural damage to buildings caused by wind speedup at isolated hills, ridges and escarpments that are abrupt changes from the general topography of the area, topographic wind effects shall be considered in the design of the building in accordance with Section R301.2.1.5.1 or in accordance with the provisions of ASCE 7. See Figure R301.2.1.5.1(1) for topographic features for wind speed-up effect.

R301.2.1.5 and Table R301.2(1) continues

R301.2.1.5 and Table R301.2(1)

Topographic Wind Effects

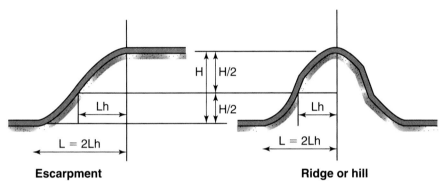

Escarpment   **Ridge or hill**

Note: H/2 determines the measurement point for Lh. L is twice Lh.

Figure R301.2.1.5.1(1) Topographic features for wind speed-up effect

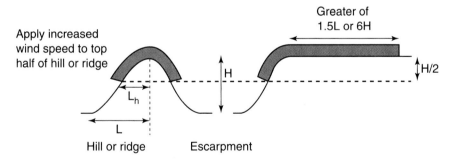

Figure R301.2.1.5.1(2) Illustration of where wind speed increase is applied on a topographic feature

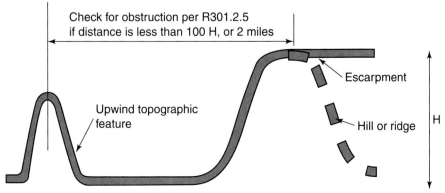

Figure R301.2.1.5.1(3) Upwind obstruction

R301.2.1.5 and Table R301.2(1)
continued

In these designated areas, topographic wind effects shall apply only to buildings sited on the top half of an isolated hill, ridge or escarpment where all of the following conditions exist:

1. The average slope of the top half of the hill, ridge or escarpment is 10 percent or greater.
2. The hill, ridge or escarpment is 60 feet (18 288 mm) or greater in height for Exposure B, 30 feet (9144 mm) or greater in height for Exposure C, and 15 feet (4572 mm) or greater in height for Exposure D.
3. The hill, ridge or escarpment is isolated or unobstructed by other topographic features of similar height in the upwind direction for a distance measured from its high point of 100 times its height or 2 miles, whichever is less. See Figure R301.2.1.5.1(3) for upwind obstruction.
4. The hill, ridge or escarpment protrudes by a factor of two or more above the height of other upwind topographic features located in any quadrant within a radius of 2 miles measured from its high point.

R301.2.1.5.1 Simplified Topographic Wind Speedup Method. As an alternative to the ASCE 7 topographic wind provisions, the provisions of R301.2.5.1 shall be permitted to be used to design for wind speedup effects, where required by R301.2.1.5.

Structures located on the top half of isolated hills, ridges, or escarpments meeting the conditions of R301.2.1.5 shall be designed for an increased basic wind speed as determined by Table R301.2.1.5.1. On the high side of an escarpment, the increased basic wind speed shall extend horizontally downwind from the edge of the escarpment 1.5 times the horizontal length of the upwind slope (1.5L) or 6 times the height of the escarpment (6H), whichever is greater. See Figure R301.2.1.5.1(2) for where wind speed increase is applied.

CHANGE SIGNIFICANCE: The provisions for topographic wind speedup effects apply only where there are historical data of structural damage from such effects, demonstrating a need for consider-

TABLE R301.2.1.5.1 Basic Wind Speed Modification for Topographic Wind Effect

Basic Wind Speed from Figure R301.2(4) (mph)	AVERAGE SLOPE OF THE TOP HALF OF HILL, RIDGE, OR ESCARPMENT						
	0.10	0.125	0.15	0.175	0.20	0.23	0.25 or greater
	Required Basic Wind Speed, Modified for Topographic Wind Speedup (mph):						
85	100	100	100	110	110	110	120
90	100	100	110	110	120	120	120
100	110	120	120	130	130	130	140
110	120	130	130	140	140	150	150
120	140	140	150	150	N/A	N/A	N/A
130	150	N/A	N/A	N/A	N/A	N/A	N/A

For SI: 1 mile per hour = 0.447 m/s

TABLE R301.2(1)   **Climatic and Geographic Design Criteria**

Ground Snow Load	Wind Design			Subject to Damage From			Winter Design Temp[e]	Ice Barrier Under-layment Required[h]	Flood Hazards[g]	Air Freezing Index[i]	Mean Annual Temp[j]
	Speed[d] (mph)	Topo-graphic Effects[k]	Seismic Design Category[f]	Weather-ing[a]	Frost line depth[b]	Termite[c]					

For SI: 1 pound per square foot = 0.0479 kPa, 1 mile per hour = 0.447 m/s.

a. through f. (No change to text.)

g. The jurisdiction shall fill in this part of the table with (a) the date of the jurisdiction's entry into the National Flood Insurance Program (date of adoption of the first code or ordinance for management of flood hazard areas), (b) the date(s) of the <u>Flood Insurance Study and (c) the panel numbers and dates</u> of all currently effective FIRM<u>s</u> and FBFM<u>s</u>, or other flood hazard map adopted by the ~~community~~ <u>authority having jurisdiction</u>, as ~~may be~~ amended.

h. through j. (No change to text.)

k. <u>In accordance with Section R301.2.1.5, where there is local historical data documenting structural damage to buildings due to topographic wind speed-up effects, the jurisdiction shall fill in this part of the table with "Yes." Otherwise, the jurisdiction shall indicate "No" in this part of the table.</u>

ation. These circumstances are most likely to occur in areas of the Pacific Northwest where there are dramatic changes in ground topography. During the code adoption process, the jurisdiction must fill in the climatic and geographical design criteria in Table R301.2(1), which now includes a column for topographic wind effects. The jurisdiction places "yes" in this table if applicable documentation of structural damage exists. Otherwise, Section R301.2.1.5 does not apply.

Buildings located on the top half of an isolated hill, ridge, or escarpment in a designated area must comply with the topographic wind speedup requirements when all of four criteria are met. These criteria relate to the height and slope of the topographic element and its location relative to other topographic features. When required, building design considering the effects of topographic wind speedup may be accomplished in one of two ways. The first is an engineered design in accordance with ASCE 7. The second is the simplified method to design for an increased basic wind speed based on the slope of the topographic feature in accordance with Table R301.2.1.5.1.

R301.2.2
Seismic Provisions

CHANGE TYPE: Clarification

CHANGE SUMMARY: Reorganization of the seismic provisions clarifies the design application within each seismic design category.

Design criteria based on Seismic Design Category (SDC)

SDC A & B All Buildings	SDC C One- and two-family dwellings	SDC C Townhouses	SDC D0, D1, & D2 All Buildings	SDC E All Buildings
No seismic requirements		Seismic provisions throughout IRC apply	Seismic provisions throughout IRC apply	Engineered design in accordance with IBC is required* *unless reclassified to lower SDC

Limitations on weights of materials

Component	Maximum average dead load
Roof/ceiling	10 psf
Roof/ceiling when wall bracing is increased per Table R301.2.2.2.1	25 psf
Floor	15 psf

Walls above grade

Exterior light frame wood	15 psf
Exterior cold-formed steel	14 psf
Interior light frame wood	10 psf
Interior cold-formed steel	5 psf
8-inch masonry	80 psf
6-inch concrete	85 psf
Structural insulated panel (SIP)	10 psf
Light frame walls with stone or masonry veneer	Sections R702.1 and R703
Fireplaces and chimneys	Chapter 10

Additional requirements

Masonry construction	R606.12

Height limitations

Type of building	Stories above grade
Wood frame	3
Cold-formed steel frame	2
Structural insulated panel (SIP)	Not permitted

Additional requirements

Masonry construction	R606.12
Concrete construction	R611 and R612
Cold-formed steel framing	AISI S230
Masonry chimneys	R1003.3 and R1103.4
Anchorage of water heaters	M1307.2

Seismic provisions

2009 CODE: R301.2.2 Seismic Provisions. The seismic provisions of this code shall apply to buildings constructed in Seismic Design Categories C, D_0, D_1 and D_2, as determined in accordance with this section. ~~Buildings in Seismic Design Category E shall be designed in accordance with the *International Building Code*, except when the seismic design category is reclassified to a lower seismic design category in accordance with Section R301.2.2.1.~~

> **Exception:** Detached one- and two-family dwellings located in Seismic Design Category C are exempt from the seismic requirements of this code.

~~The weight and irregularity limitations of Section R301.2.2.2 shall apply to buildings in all seismic design categories regulated by the seismic provisions of this code. Buildings in Seismic Design Category C shall be constructed in accordance with the additional requirements of Section 301.2.2.3. Buildings in Seismic Design Categories D_0, D_1 and D_2 shall be constructed in accordance with the additional requirements of Section R301.2.2.4.~~

R301.2.2.2 Seismic Limitations. ~~The following limitations apply to buildings in all Seismic Design Categories regulated by the seismic provisions of this code.~~

~~**R301.2.2.3**~~ <u>**R301.2.2.2**</u> **Seismic Design Category C.** Structures assigned to Seismic Design Category C shall conform to the requirements of this section.

R301.2.2.2.1 Weights of Materials. Average dead loads shall not exceed 15 pounds per square foot (720 Pa) for the combined roof and ceiling assemblies (on a horizontal projection) or 10 pounds per square foot (480 Pa) for floor assemblies, except as further limited by Section R301.2.2. Dead loads for walls above grade shall not exceed:

1. Fifteen pounds per square foot (720 Pa) for exterior light-frame wood walls.

2. Fourteen pounds per square foot (670 Pa) for exterior light-frame cold-formed steel walls.

3. Ten pounds per square foot (480 Pa) for interior light-frame wood walls.

4. Five pounds per square foot (240 Pa) for interior light-frame cold-formed steel walls.

5. Eighty pounds per square foot (3830 Pa) for 8-inch-thick (203 mm) masonry walls.

6. Eighty-five pounds per square foot (4070 Pa) for 6-inch-thick (152 mm) concrete walls.

7. <u>Ten pounds per square foot (480 Pa) for SIP walls.</u>

R301.2.2 continues

R301.2.2 continued

TABLE R301.2.2.2.1 Wall Bracing Adjustment Factors by Roof Covering Dead Load[a]

Wall Supporting	Roof/Ceiling Dead Load	Roof/Ceiling Dead Load
	15 psf or less	25 psf
Roof only	1.0	1.2
Roof plus one ~~story~~ or two stories	1.0	1.1

For SI: 1 pound per square foot = 0.049 kPa.
a. Linear interpolation shall be permitted.

Exceptions: (No change to text.)

~~R301.2.2.4~~ R301.2.2.3 Seismic Design Categories D$_0$, D$_1$ and D$_2$. Structures assigned to Seismic Design Categories D$_0$, D$_1$ and D$_2$ shall conform to the requirements for Seismic Design Category C and the additional requirements of this section.

~~R301.2.2.4.1~~ R301.2.2.3.1 Height Limitations. Wood framed buildings shall be limited to three stories above grade or the limits given in Table R602.10.1.2(2). Cold-formed steel framed buildings shall be limited to less than or equal to ~~two~~ three stories above grade in accordance with ~~COFS/PM~~ AISI S230. Mezzanines as defined in Section R202 shall not be considered as stories. Structural insulated panel buildings shall be limited to two stories above grade.

~~R301.2.2.4.5~~ R301.2.2.3.5 Cold-formed Steel Framing in Seismic Design Categories D$_0$, D$_1$ and D$_2$. In Seismic Design Categories D$_0$, D$_1$ and D$_2$ in addition to the requirements of this code, cold-formed steel framing shall comply with the requirements of ~~COFS/PM~~ AISI S230.

R301.2.2.3.6 Masonry Chimneys. Masonry chimneys shall be reinforced and anchored to the building in accordance with Sections R1003.3 and R1003.4.

R301.2.2.3.7 Anchorage of Water Heaters. Water heaters shall be anchored against movement and overturning in accordance with Section M1307.2.

Because these code changes completely reorganized Section R301.2.2, the entire code text is too extensive to be included here. Refer to Code Changes RB34-06/07, RB44-06/07, and RB179–07/08 in the *2009 IRC Code Changes Resource Collection* for the complete text and history of the code changes.

CHANGE SIGNIFICANCE: The seismic provisions have been rearranged in a sequential order to clarify when they apply. The general statement that there are specific requirements applicable to buildings in Seismic Design Category (SDC) C, D$_0$, D$_1$, and D$_2$ remains at the beginning of Section R301.2.2, indicating that there are no additional requirements in SDC A or B. The exemption for one- and two-family dwellings in SDC C also remains at the beginning. The provisions for SDC E are moved to the end of the section. One of the confusing issues of the previous code was the second paragraph, which sent the user to provisions that applied to all SDCs, followed by reference to SDC C requirements and SDC D_0, D_1, and D_2 requirements. This paragraph and the section applying certain requirements to all SDCs have been removed. All requirements that apply to buildings located in SDC C now immediately follow the charging statement in R301.2.2 and the determination of Seismic Design Category in R301.2.2.1. Appearing next are the SDC D_0, D_1, and D_2 requirements, which include all SDC C provisions and a number of additional limitations. Height limitations have been added for structural insulated panels (SIPs), which are now recognized in the prescriptive provisions of the IRC. Masonry chimney and water heater anchorage requirements have been added to the list under SDC D_0, D_1, and D_2 to call attention to the corresponding provisions in Sections R1003 and M1307.

R301.2.3

Snow Loads

CHANGE TYPE: Modification

CHANGE SUMMARY: Structural insulated panels (SIPs) have been added to the list of approved prescriptive construction methods that are limited to a maximum ground snow load of 70 psf.

2009 CODE: R301.2.3 Snow Loads. Wood framed construction, cold-formed steel framed construction, ~~and~~ masonry and concrete construction, and structural insulated panel construction in regions with ground snow loads 70 pounds per square foot (3.35 kPa) or less, shall be in accordance with Chapters 5, 6 and 8. Buildings in regions with ground snow loads greater than 70 pounds per square foot (3.35 kPa) shall be designed in accordance with accepted engineering practice.

CHANGE SIGNIFICANCE: The IRC now includes prescriptive requirements for construction with SIPs. As with conventional wood and cold-formed steel framed construction, the code places limits on prescriptive SIP construction for wind, seismic, and snow load design criteria. Buildings located in areas where the ground snow load exceeds 70 psf must be designed in accordance with accepted engineering practice, a requirement that now includes SIP construction.

SIP construction is limited to:
10 ft. wall height
2 stories
70 psf ground snow load

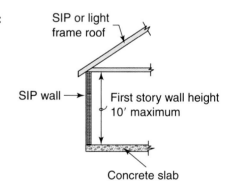

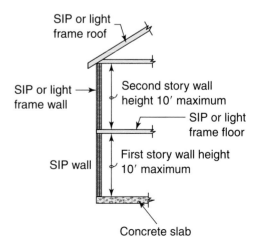

Snow load limits on structural insulated panel (SIP) construction

CHANGE TYPE: Modification

CHANGE SUMMARY: In the prescriptive wood framing provisions, floor framing is now permitted to exceed 16 inches in height, provided the overall story height is not exceeded. Structural insulated panel (SIP) bearing walls are limited to 10 feet in height.

2009 CODE: R301.3 Story Height. Buildings constructed in accordance with these provisions shall be limited to story heights of not more than the following:

1. For wood wall framing, the laterally unsupported bearing wall stud height permitted by Table R602.3(5) plus a height of floor framing not to exceed 16 inches.

Exception: For wood framed wall buildings with bracing in accordance with Tables R602.10.1.2(1) and R602.10.1.2(2), the wall stud clear height used to determine the maximum permitted story height may be increased to 12 feet (3658 mm) without requiring an engineered design for the building wind and seismic force resisting systems provided that the length of bracing required by Table R602.10.1.2(1) is increased by multiplying by a factor of 1.10 and the length of bracing required by Table R602.10.1.2(2) is increased by multiplying by a factor of 1.20. Wall studs are still subject to the requirements of this section.

2. For steel wall framing, a stud height of 10 feet (3048 mm), plus a height of floor framing not to exceed 16 inches (406 mm).

3. For masonry walls, a maximum bearing wall clear height of 12 feet (3658 mm) plus a height of floor framing not to exceed 16 inches (406 mm).

R301.3 continues

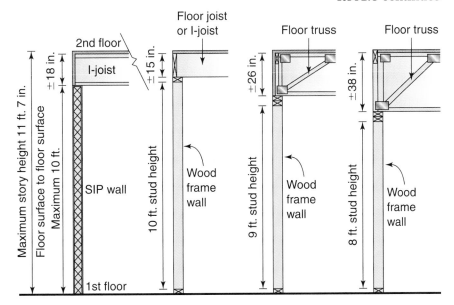

Story height

R301.3 continued

Exception: An additional 8 feet (2438 mm) is permitted for gable end walls.

4. For insulating concrete form walls, the maximum bearing wall height per story as permitted by Section R611 tables plus a height of floor framing not to exceed 16 inches (406 mm).

5. For structural insulated panel (SIP) walls, the maximum bearing wall height per story as permitted by Section R614 tables shall not exceed 10 feet (3048 mm) plus a height of floor framing not to exceed 16 inches (406 mm).

Individual walls or wall studs shall be permitted to exceed these limits as permitted by Chapter 6 provisions, provided story heights are not exceeded. Floor framing height shall be permitted to exceed these limits provided the story height does not exceed 11 feet 7 inches (3531 mm). An engineered design shall be provided for the wall or wall framing members when they exceed the limits of Chapter 6. Where the story height limits are exceeded, an engineered design shall be provided in accordance with the *International Building Code* for the overall wind and seismic force resisting systems.

CHANGE SIGNIFICANCE: When using the prescriptive wood framing provisions in Chapter 6, the code previously limited story height to the sum of the tabular value for stud wall height plus 16 inches for the floor framing height. The intent was to limit story height, measured from the finished floor surface of one story to the finished floor surface of the next story, but the provision in effect limited the floor framing height without technical justification. Whereas Chapter 6 limits the wall stud height, there is no such height limitation in Chapter 5 for floor framing. This section has been interpreted as applying to any wood floor framing method, not just solid lumber joists. Wood I-joists and open web floor trusses can and often do exceed 16 inches in height. The new language permits floor framing to exceed the 16-inch height limit, provided the story height does not exceed 11 feet 7 inches. Therefore, where 10-foot studs are used, the maximum floor framing height is approximately 16 inches, but may be increased when shorter studs are used.

The code also now includes prescriptive provisions for SIP construction. The height of SIP bearing walls is limited to 10 feet. As with conventional framed walls, the floor framing is permitted to exceed 16 inches in height as long as the story height does not exceed 11 feet 7 inches.

CHANGE TYPE: Modification

CHANGE SUMMARY: The definitions for deck and balcony have been removed, and the minimum uniform live load for balconies has been lowered from 60 psf to 40 psf to be consistent with decks. The criteria for determining a limited attic storage area now considers the required depth of the insulation relative to the truss bottom chord depth. Habitable attics and attics served with fixed stairs have been added to the table and have a minimum uniform live load of 30 psf.

Table R301.5 continues

Table R301.5

Minimum Uniformly Distributed Live Loads

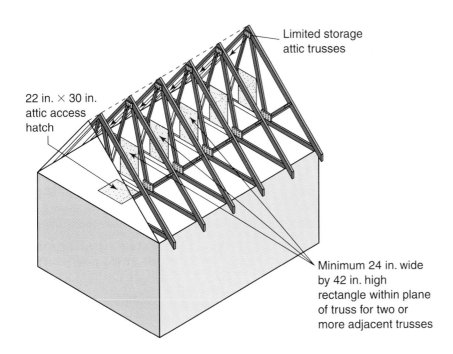

Limited storage attic trusses

22 in. × 30 in. attic access hatch

Minimum 24 in. wide by 42 in. high rectangle within plane of truss for two or more adjacent trusses

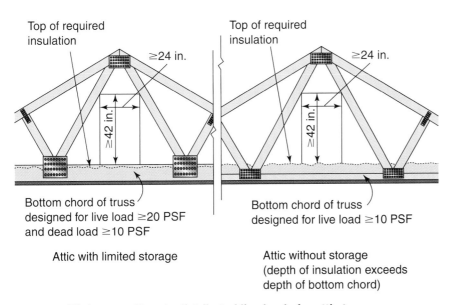

Top of required insulation

≥24 in.

≥42 in.

Bottom chord of truss designed for live load ≥20 PSF and dead load ≥10 PSF

Attic with limited storage

Top of required insulation

≥24 in.

≥42 in.

Bottom chord of truss designed for live load ≥10 PSF

Attic without storage (depth of insulation exceeds depth of bottom chord)

Minimum uniformly distributed live loads for attic trusses

Table R301.5 continued

TABLE R301.5   Minimum Uniformly Distributed Live Loads (in pounds per square foot)

Use	Live Load
~~Attics with limited storage~~[b, g, h]	~~20~~
Attics without storage[b]	10
Attics with limited storage[b, g]	20
Habitable attics and attics served with fixed stairs	30
Balconies (exterior) and decks[e]	40
~~Exterior balconies~~	~~60~~
Fire escapes	40
Guardrails and handrails[d]	200[h]
Guardrail in-fill components[f]	50[h]
Passenger vehicle garages[a]	50[a]
Rooms other than sleeping rooms	40
Sleeping rooms	30
Stairs	40[c]

For SI: 1 pound per square foot = 0.0479 kPa, 1 square inch = 645 mm^2, 1 pound = 4.45 N.

a. Elevated garage floors shall be capable of supporting a 2,000-pound load applied over a 20-square-inch area.

b. Attics without storage are those where the maximum clear height between joist and rafter is less than 42 inches, or where there are not two or more adjacent trusses with the same web configuration capable of containing a rectangle 42 inches high by 2 feet wide, or greater, located within the plane of the truss. For attics without storage, this live load need not be assumed to act concurrently with any other live load requirements.

c. Individual stair treads shall be designed for the uniformly distributed live load or a 300-pound concentrated load acting over an area of 4 square inches, whichever produces the greater stresses.

d. A single concentrated load applied in any direction at any point along the top.

e. See Section R502.2.2 for decks attached to exterior walls.

f. Guard in-fill components (all those except the handrail), balusters, and panel fillers shall be designed to withstand a horizontally applied normal load of 50 pounds on an area equal to 1 square foot. This load need not be assumed to act concurrently with any other live load requirement.

g. For attics with limited storage and constructed with trusses, this live load need be applied only to those portions of the bottom chord where there are two or more adjacent trusses with the same web configuration capable of containing a rectangle 42 inches high or greater by 2 feet wide or greater, located within the plane of the truss. The rectangle shall fit between the top of the bottom chord and the bottom of any other truss member, provided that each of the following criteria is met:

   1. The attic area is accessible by a pull-down stairway or framed opening in accordance with Section R807.1.
   2. The truss has a bottom chord pitch less than 2:12.
   3. Required insulation depth is less than the bottom chord member depth.

The bottom chords of trusses meeting the above criteria for limited storage shall be designed for the greater of the actual imposed dead load or 10 psf, uniformly distributed over the entire span.

~~h. Attic spaces served by a fixed stair shall be designed to support the minimum live load specified for sleeping rooms.~~

~~i.~~ h. Glazing used in handrail assemblies and guards shall be designed with a safety factor of 4. The safety factor shall be applied to each of the concentrated loads applied to the top of the rail, and to the load on the in-fill components. These loads shall be determined independent of one another, and loads are assumed not to occur with any other live load.

2009 CODE

Section R202, Definitions
Balcony, Exterior. ~~An exterior floor projecting from and supported by a structure without additional independent supports.~~

Deck. ~~An exterior floor system supported on at least two opposing sides by an adjoining structure and/or posts, piers, or other independent supports.~~

CHANGE SIGNIFICANCE: In general, decks and balconies have different methods of support—a deck relies on the structure and independent posts, while a balcony is supported entirely by the structure—but both serve the same function. It follows that the anticipated live load for each would be the same. The rationale for the higher live load requirement for balconies, as has appeared in previous editions of the IRC, has been that balconies do not have redundancy. In other words, a single point of failure could presumably cause the entire balcony to fail. Decks, on the other hand, have multiple points of support and are considered by some to therefore have a higher factor of safety. There is insufficient available data on deck and balcony failures to conclude that the higher live load requirements for balconies are justified. When a deck failure does occur, it is typically at the ledger connection to the structure or because of faulty post connections at the footing or beam. In addition, any concerns that cantilevers are less redundant than supported structures should be addressed by increasing the safety factor of the design, not by increasing the minimum live load. Loads are not affected by the method of support. The minimum live load of 40 psf is appropriate for balconies and decks regulated by the IRC because it matches the requirements for living areas of the dwelling. The definitions for balconies and decks are no longer needed to distinguish the different live load requirements, and those definitions have been deleted.

The change to footnote *g* of Table R301.5 clarifies the application of dead load on the bottom chord in the truss design for attics with limited storage. In addition, the depth of required attic insulation is now considered in determining a limited attic storage area. If the insulation thickness exceeds the depth of the bottom chord, it is unlikely the area will be used for storage, and the higher loading requirements are not necessary.

A new defined term in the 2009 IRC, *habitable attics* are occupiable space between the uppermost floor/ceiling assembly and the roof assembly. They must meet the minimum room size and ceiling height requirements for habitable spaces but are not considered a story. Habitable attics require a smoke detector, emergency escape and rescue opening, and a means of egress complying with Section R311. The minimum uniform live load for habitable attics is 30 psf, consistent with sleeping rooms and nonoccupied attics served by fixed stairs. Footnote *h* has been deleted, and the minimum live load for attics with fixed stairs, which includes habitable attics, has been placed in the table.

R302.1 and Table R302.1

Fire-resistant Construction at Exterior Walls

CHANGE TYPE: Modification

CHANGE SUMMARY: Section R302 has been renamed *Fire-resistant Construction* and pulls in related provisions from sections on separations, penetrations, and other fire-resistance requirements so that they reside in one section and can be more easily located. Exterior walls requiring a one-hour fire-resistance rating due to fire separation distance must now meet the requirements of ASTM E 119 or UL 263. Fire separation distance requirements no longer apply to buildings on the same lot. Changes to Table R302.1 clarify the application of the fire separation distance requirements.

2009 CODE:
Section R302

~~**Exterior Wall Location**~~ **Fire-resistant Construction**

R302.1 Exterior Walls. Construction, projections, openings and penetrations of exterior walls of dwellings and accessory buildings shall comply with Table R302.1. ~~These provisions shall not apply to walls, projections, openings or penetrations in walls that are perpendicular to the line used to determine the fire separation distance. Projections beyond the exterior wall shall not extend more than 12 inches (305 mm) into the areas where openings are prohibited.~~

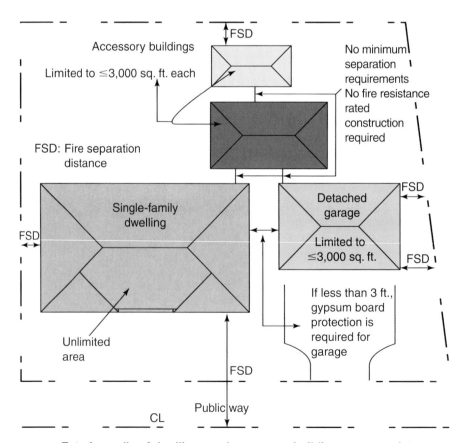

Exterior walls of dwellings and accessory buildings on same lot

Exceptions:

<u>**1.** Walls, projections, openings, or penetrations in walls perpendicular to the line used to determine the fire separation distance.</u>

<u>**2.** Walls of dwellings and accessory structures located on the same lot.</u>

~~1~~**3.** Detached tool sheds and storage sheds, playhouses and similar structures exempted from permits are not required to provide wall protection based on location on the lot. Projections beyond the exterior wall shall not extend over the lot line.

~~2~~**4.** Detached garages accessory to a dwelling located within 2 feet (610 mm) of a lot line are permitted to have roof eave projections not exceeding 4 inches (102 mm).

~~3~~**5.** Foundation vents installed in compliance with this code are permitted.

Because these code changes reorganized substantial portions of Chapter 3, placing all fire-resistance requirements into Section R302, the entire code text is too extensive to be included here. Refer to Code Changes RB53-06/07, RB56-06/07, RB21–07/08, RB23–07/08, and RB24–07/08 in the *2009 IRC Code Changes Resource Collection* for the complete text and history of the code changes related to Section R302.

CHANGE SIGNIFICANCE: In an effort to make the code more user friendly and to follow a similar format as that found in Chapter 7 of the IBC, Chapter 3 has been reorganized to place all fire-resistance provisions into Section R302. Changes to the text of the various sec-

R302.1 and Table R302.1 continues

TABLE R302.1   **Exterior Walls**

Exterior Wall Element		Minimum Fire-Resistance Rating	Minimum Fire Separation Distance
Walls	(Fire-resistance rated)	1 hour <u>tested in accordance with ASTM E 119 or UL 263</u> with exposure from both sides	$\theta \leq 5$ feet
	(Not fire-resistance rated)	0 hours	≥ 5 feet
Projections	(Fire-resistance rated)	1 hour on the underside	≥ 2 feet <u>to 5 feet</u>
	(Not fire-resistance rated)	0 hours	5 feet
Openings <u>in walls</u>	Not allowed	N/A	< 3 feet
	25% maximum of wall area	0 hours	3 feet
	Unlimited	0 hours	5 feet
Penetrations	All	Comply with Section R302.4	< 5 feet
		None required	5 feet

R302.1 and Table R302.1 continued

tions that are brought into R302 are minor and editorial. Section R302 now contains provisions for dwelling unit separations, penetrations, dwelling-garage separation, flame spread and smoke-developed indices of finish materials, fire blocking, and draftstopping.

Exterior walls with a fire separation distance of less than 5 feet from a lot line require a one-hour fire-resistance rating with exposure from both sides. Though building officials typically relied on rated assemblies meeting nationally recognized test standards to verify compliance, this section of the code was silent regarding approved test standards. The added language in Table R302.1 requires that the one-hour wall assemblies meet the test requirements of ASTM E 119 *Standard Test Methods for Fire Tests of Building Construction and Materials* or UL 263 *Fire Tests of Building Construction and Materials.*

The significant change in Section R302.1 of the 2009 IRC intends to clarify that no separation distance or fire-resistance rating is required between detached structures on the same lot. Proponents reasoned that since the IRC does not limit the area of dwellings, it should not impose fire-resistance requirements on detached buildings. Separate buildings on the same residential lot could reasonably be considered one building, similar to provisions in the IBC when the aggregate area of multiple buildings falls within the allowable area limitations. With no limits on dwelling area and an area limitation of 3000 square feet in place for accessory structures in the IRC, the exemption for fire separation distance between buildings on the same lot is not considered to create any undue risk to the occupants. There has previously been inconsistent application in determining the fire resistance and opening protection requirements for the opposing walls of a dwelling and a detached accessory structure. In many cases, officials have referenced the definition of fire separation distance in establishing an imaginary line between the two buildings in making a determination. There has been no clear charging statement in the code to require an imaginary line, however, leading many to determine that fire-resistant construction was not required even when the buildings were in close proximity. The new exception intends to resolve this issue. Clearly, the intent is not to waive all fire separation requirements for buildings on the same lot—for example, prescribed distances to lot lines—but only to address the walls of the detached buildings that face each other. This new exception also does not intend to apply to townhouses, which may be constructed with or without property lines between them but in all cases require a fire-resistant separation. It is important to note that the separation requirements (i.e., ½-inch gypsum board on the garage side) for a detached garage less than 3 feet from a dwelling in Sections R302.5 and R302.6 are still in effect.

Minor changes to Table R302.1 clarify the minimum separation distance measurements and emphasize that the opening requirements apply to openings only in walls, not those in soffits or roofs.

CHANGE TYPE: Modification

CHANGE SUMMARY: The dwelling unit separation provisions have been relocated from Section R317 to Section R302. The code now recognizes UL 263 as an equivalent test standard to ASTM E 119 for fire resistance. Both test standards are now referenced as meeting the test requirements for the required fire-resistance rating between dwelling units of two-family dwellings and townhouses. A common 1-hour fire-resistance-rated wall satisfies the townhouse separation requirement. New language clarifies that wall assemblies separating two-family dwellings must begin at the foundation.

2009 CODE: ~~R317.2~~ R302.2 Townhouses. Each townhouse shall be considered a separate building and shall be separated by fire-resistance-rated wall assemblies meeting the requirements of Section ~~R302~~ R302.1 for exterior walls.

> **Exception:** A common ~~2~~ 1-hour fire-resistance-rated wall <u>assembly tested in accordance with ASTM E 119 or UL 263</u> is permitted for townhouses if such walls do not contain plumbing or mechanical equipment, ducts or vents in the cavity of the common wall. <u>The wall shall be rated for fire exposure from both sides and shall extend to and be tight against exterior walls and the underside of the roof sheathing.</u> Electrical

R302.2 and R302.3
Dwelling Unit Separation

R302.2 and R302.3 continues

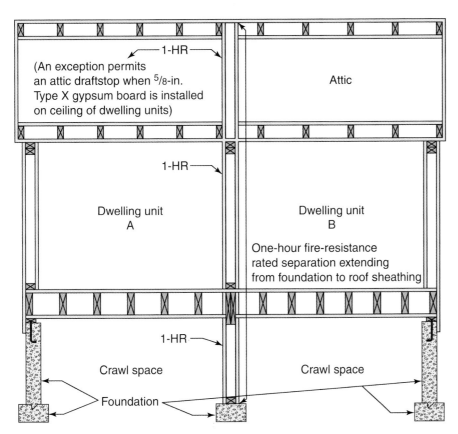

Two-family dwelling separation wall

R302.2 and R302.3 continued

installations shall be installed in accordance with Chapters 34 through 43. Penetrations ~~of electrical outlet boxes~~ shall be in accordance with Section ~~R317.3~~ <u>R302.4</u>.

~~R317.2.4~~ <u>R302.2.4</u> Structural Independence.   Each individual townhouse shall be structurally independent.

Exceptions:
1. Foundations supporting exterior walls or common walls.
2. Structural roof and wall sheathing from each unit may fasten to the common wall framing.
3. Nonstructural wall <u>and roof</u> coverings.
4. Flashing at termination of roof covering over common wall.
5. Townhouses separated by a common ~~2~~ <u>1</u>-hour fire-resistance-rated wall as provided in Section ~~R317.2~~ <u>R302.2</u>.

~~Section R317 Dwelling Unit Separation~~
~~R317.1~~ <u>R302.3</u> Two-Family Dwellings.   Dwelling units in two-family dwellings shall be separated from each other by wall and/or floor assemblies having not less than a 1-hour fire-resistance rating when tested in accordance with ASTM E 119 <u>or UL 263</u>. Fire-resistance-rated floor-ceiling and wall assemblies shall extend to and be tight against the exterior wall, and wall assemblies shall extend <u>from the foundation</u> to the underside of the roof sheathing.

Exceptions:
1. A fire-resistance rating of ½ hour shall be permitted in buildings equipped throughout with an automatic sprinkler system installed in accordance with NFPA 13.
2. Wall assemblies need not extend through attic spaces when the ceiling is protected by not less than $\frac{5}{8}$-inch (15.9 mm) Type X gypsum board and an attic draft stop constructed as specified in Section R502.12.1 is provided above and along the wall assembly separating the dwellings. The structural framing supporting the ceiling shall also be protected by not less than ½-inch (12.7 mm) gypsum board or equivalent.

CHANGE SIGNIFICANCE: In the consolidation of the fire-resistant construction provisions, the dwelling unit separation requirements for townhouses and two-family dwellings in Sections R317.1 and R317.2 of the 2006 IRC have been relocated to Section R302.

The fire-resistance rating for the common wall between townhouses has been reduced from two hours to one hour. This change is coupled with new requirements for an automatic fire sprinkler system to be installed in all townhouses (as discussed later in this publication). Jurisdictions in many parts of the country have permitted townhouse separation with a one-hour fire-resistance rating with the installation of a sprinkler system. Based on satisfactory results of these installations, this change was put forth to address a reasonable level of fire protection and to limit the impact on affordable housing.

Fire-resistance test standards ASTM E 119 and UL 263 have identical specifications for test apparatus and test procedures resulting in identical test results. UL 263 has been added as an approved test method for the fire-resistance-rated separations between two-family dwellings. The inclusion of this alternate test method provides the authority having jurisdiction with the flexibility to accept listed and labeled products evaluated in accordance with ASTM E 119 or UL 263. Although the code previously listed ASTM E 119 as a test standard for the fire-resistance-rated wall assemblies for two-family dwellings in R317.1, the IRC did not identify testing methodology to determine the fire-resistance rating for separations between townhouses in Section R317.2 or for exterior walls in Section R302 related to fire separation distance (discussed previously in this publication). The code now requires these assemblies to be tested in accordance with ASTM E 119 or UL 263, giving the building official specific direction for verifying the fire-resistant ratings.

In order to ensure a complete separation including basement and crawl space areas, this code change also clarifies that fire-resistance-rated wall assemblies for separation of two-family dwellings must be continuous from the foundation to the roof.

R302.4

Rated Penetrations for Dwelling Unit Separation

CHANGE TYPE: Modification

CHANGE SUMMARY: The rated penetration provisions for dwelling unit separation have been relocated from Section R317 to Section R302. Editorial changes clarify the exceptions for metal pipe and conduit penetrations. Modification of the provisions for electrical boxes more accurately represents accepted practices for preserving the fire-resistance rating of walls.

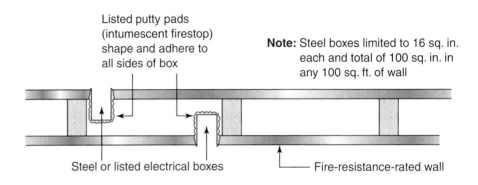

Listed putty pads (intumescent firestop) shape and adhere to all sides of box

Note: Steel boxes limited to 16 sq. in. each and total of 100 sq. in. in any 100 sq. ft. of wall

Steel or listed electrical boxes

Fire-resistance-rated wall

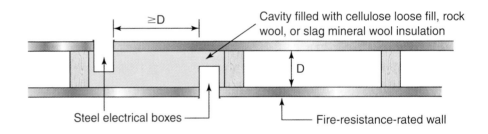

$\geq D$

Cavity filled with cellulose loose fill, rock wool, or slag mineral wool insulation

D

Steel electrical boxes

Fire-resistance-rated wall

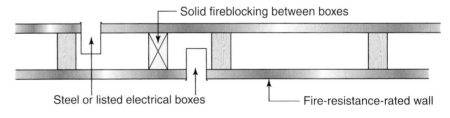

Solid fireblocking between boxes

Steel or listed electrical boxes

Fire-resistance-rated wall

*Note: Separation distance for listed boxes is determined by the listing

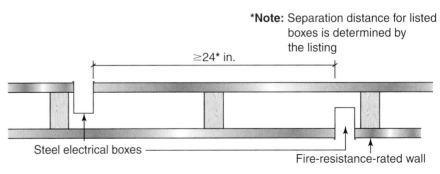

$\geq 24*$ in.

Steel electrical boxes

Fire-resistance-rated wall

Electrical box penetrations on opposite sides of rated wall assembly

2009 CODE: R317.3 R302.4 Dwelling Unit Rated Penetrations.
Penetrations of wall or floor/ceiling assemblies required to be fire-resistance rated in accordance with Section R317.1 or R317.2 R302.2 or R302.3 shall be protected in accordance with this section.

R317.3.1 R302.4.1 Through Penetrations.   Through penetrations of fire-resistance-rated wall or floor assemblies shall comply with Section R317.3.1.1 or R317.3.1.2 R302.4.1.1 or R302.4.1.2.

> **Exception:** Where the penetrating items are steel, ferrous or copper pipes, tubes or conduits, the annular space shall be protected as follows:
>
> 1. In concrete or masonry wall or floor assemblies where the penetrating item is a maximum 6 inches (152 mm) nominal diameter and the area of the opening through the wall does not exceed 144 square inches (92 900 mm²), concrete, grout or mortar is shall be permitted where installed to the full thickness of the wall or floor assembly or the thickness required to maintain the fire-resistance rating, provided:
>
>> 1.1. The nominal diameter of the penetrating item is a maximum of 6 inches (152 mm); and
>> 1.2. The area of the opening through the wall does not exceed 144 square inches (92 900 mm²).
>
> 2. The material used to fill the annular space shall prevent the passage of flame and hot gases sufficient to ignite cotton waste where subjected to ASTM E 119 or UL 263 time temperature fire conditions under a minimum positive pressure differential of 0.01 inch of water (3 Pa) at the location of the penetration for the time period equivalent to the fire-resistance rating of the construction penetrated.

R317.3.1.1 R302.4.1.1 Fire-Resistance-Rated Assembly. (No change to text.)

R317.3.1.2 R302.4.1.2 Penetration Firestop System. (No change to text.)

R317.3.2 R302.4.2 Membrane Penetrations.   Membrane penetrations shall comply with Section R317.3.1 R302.4.1. Where walls are required to have a fire-resistance rating, recessed fixtures shall be so installed such that the required fire resistance rating will not be reduced.

> **Exceptions:**
> 1. Membrane penetrations of maximum 2-hour fire-resistance-rated walls and partitions by steel electrical boxes that do not exceed 16 square inches (0.0103 m²) in area, provided the aggregate area of the openings through the membrane does not exceed 100 square inches (0.0645 m²) in any 100 square feet (9.29 m²) of wall area. The annular space between the wall membrane and the box shall not exceed ⅛ inch (3.1 mm). Such

R302.4 continues

R302.4 continued

boxes on opposite sides of the wall shall be separated ~~as follows~~ by one of the following:

1.1. By a horizontal distance of not less than 24 inches (610 mm) ~~except at walls or partitions constructed using parallel rows of studs or staggered studs~~ where the wall or partition is constructed with individual noncommunicating stud cavities;

1.2. By a horizontal distance of not less than the depth of the wall cavity when the wall cavity is filled with cellulose loose-fill, rockwool or slag mineral wool insulation;

1.3. By solid fire blocking in accordance with Section ~~R602.8.1~~ R302.11;

1.4. By protecting both boxes with listed putty pads; or

1.5. By other listed materials and methods.

2. Membrane penetrations by listed electrical boxes of any materials provided the boxes have been tested for use in fire-resistance-rated assemblies and are installed in accordance with the instructions included in the listing. The annular space between the wall membrane and the box shall not exceed $\frac{1}{8}$ inch (3.1 mm) unless listed otherwise. Such boxes on opposite sides of the wall shall be separated ~~as follows~~ by one of the following:

2.1. By ~~a~~ the horizontal distance ~~of not less than 24 inches (610 mm) except at walls or partitions constructed using parallel rows of studs or staggered studs~~ specified in the listing of the electrical boxes;

2.2. By solid fire blocking in accordance with Section ~~R602.8~~ R302.11;

2.3. By protecting both boxes with listed putty pads; or

2.4. By other listed materials and methods.

3. The annular space created by the penetration of a fire sprinkler provided it is covered by a metal escutcheon plate.

CHANGE SIGNIFICANCE: In the consolidation of the fire-resistant construction provisions, the rated penetration requirements for dwelling unit separations in Section R317.3 of the 2006 IRC have been relocated to Section R302.4.

The exception to Section R302.4.1 permits penetrating items of specified metal pipe or conduit in two instances in lieu of a listed assembly or penetration firestop system. In the first instance, for concrete and masonry assemblies, the firestopping materials may be concrete, grout, or masonry. The conditions for maximum pipe diameter and area of opening of the penetrations are now listed separately under this type of installation, so they are more readily understandable to the code user. The phrase "shall be permitted" is also inserted to provide uniformity in the mandatory language. In the second instance, for filling the annular space by materials complying with fire-resistance testing, the code now recognizes UL 263 as an equivalent test standard to ASTM E 119.

The general rule for membrane penetration by recessed fixtures requires that the fire-resistance rating of the wall assembly be maintained. The exceptions apply to two types of electrical boxes—steel boxes and listed boxes of any material. The prescribed methods for preserving the integrity of the fire-resistance-rated wall have been revised to more accurately describe electrical box installation requirements. For steel boxes, the 24-inch offset of boxes on opposite sides of the wall now applies only when the wall is constructed of individual stud cavities, which do not communicate with adjacent stud cavities. This change recognizes that construction methods other than parallel or staggered studs—for example, installation of resilient channels on studs—would allow products of combustion to bypass the stud separating the two boxes and reach the electrical box penetration on the opposite side of the wall. The 24-inch separation rule has been eliminated as an option for listed electrical boxes. Instead, listed boxes must be separated by the horizontal distance specified in the listing.

R302.5

Garage Openings and Penetrations

CHANGE TYPE: Modification

CHANGE SUMMARY: The dwelling/garage separation provisions in Sections R309.1 and R309.2 of the 2006 IRC have been relocated to Section R302 with the other fire-resistant construction provisions. A new charging statement has been added as Section R302.5 for clarification of the opening and penetration details. Penetration requirements now reference the fireblocking provisions that have been relocated to Section R302.11 (the fireblocking provisions formerly appeared in the wood wall framing provisions of Section R602.8).

2009 CODE: R302.5 Dwelling/Garage Opening/Penetration Protection. Openings and penetrations through the walls or ceilings separating the dwelling from the garage shall be in accordance with Sections R302.5.1 through R302.5.3.

~~R309.1~~ **R302.5.1 Opening Protection.** Openings from a private garage directly into a room used for sleeping purposes shall not be permitted. Other openings between the garage and residence shall be equipped with solid wood doors not less than $1\frac{3}{8}$ inches (35 mm) in thickness, solid or honeycomb core steel doors not less than $1\frac{3}{8}''$ inches (35 mm) thick, or 20-minute fire-rated doors.

~~R309.1.1~~ **R302.5.2 Duct Penetration.** Ducts in the garage and ducts penetrating the walls or ceilings separating the dwelling from the garage shall be constructed of a minimum No. 26 gage (0.48 mm) sheet steel or other approved material and shall have no openings into the garage.

~~R309.1.2~~ **R302.5.3 Other Penetrations.** Penetrations through the separation required in Section R302.6 shall be protected <u>as required by Section R302.11, Item 4.</u> ~~by filling the opening around the penetrat-~~

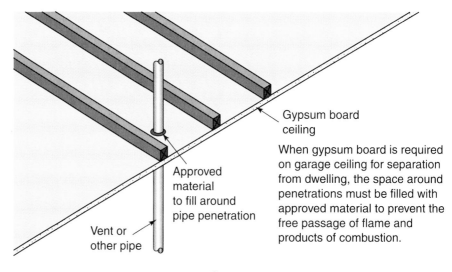

Gypsum board ceiling

When gypsum board is required on garage ceiling for separation from dwelling, the space around penetrations must be filled with approved material to prevent the free passage of flame and products of combustion.

Approved material to fill around pipe penetration

Vent or other pipe

Garage

Penetration of ceiling separating garage from dwelling

~~ing item with approved material to resist the free passage of flame and products of combustion.~~

CHANGE SIGNIFICANCE: The provisions for door openings and duct penetrations through the separation between the dwelling and the garage have not changed. Previously, penetrations other than ducts required the opening around the penetrating item to be filled with approved material to restrict passage of combustion products. Now the code specifically references Item 4 of the fireblocking provisions, which are relocated from Section R602.8 to Section R302.11, for such protection. Fireblocking is by definition the installation of building materials installed to resist the free passage of flame to other areas of the building through concealed spaces. Section R302.11 requires fireblocking to cut off all concealed draft openings and to form an effective fire barrier between stories and between a top story and the roof space. Item 4 applies only to openings at the floor and ceiling level and requires an approved material to resist the free passage of combustion products. It follows that the application for this change is to seal around openings of pipes, vents, cables, and wires penetrating the separation at the floor and ceiling level. It will not apply to ducts because duct penetrations must follow the prescribed installation methods of Section R302.5.2.

R302.6 and Table R302.6

Garage Separation

CHANGE TYPE: Clarification

CHANGE SUMMARY: The dwelling/garage separation provisions in Sections R309.1 and R309.2 of the 2006 IRC have been relocated to Section R302 with the other fire-resistant construction provisions. For clarification, the provisions requiring the application of gypsum board on the garage side of the separation from a dwelling have been placed in a new table, and the corresponding text has been deleted from Section R309.2.

2009 CODE: ~~**R309.2 Separation Required.**~~ ~~The garage shall be separated from the residence and its attic area by not less than ½-inch (12.7 mm) gypsum board applied to the garage side. Garages beneath habitable rooms shall be separated from all habitable rooms above by not less than 5/8-inch (15.9 mm) Type X gypsum board or equivalent. Where the separation is a floor-ceiling assembly, the structure supporting the separation shall also be protected by not less than 1/2-inch (12.7 mm) gypsum board or equivalent. Garages located less than 3 feet (914 mm) from a dwelling unit on the same lot shall be protected with not less than 1/2-inch (12.7 mm) gypsum board applied to the interior side of exterior walls that are within this area. Openings in these walls shall be regulated by Section R309.1. This provision does not apply to garage walls that are perpendicular to the adjacent dwelling unit wall.~~

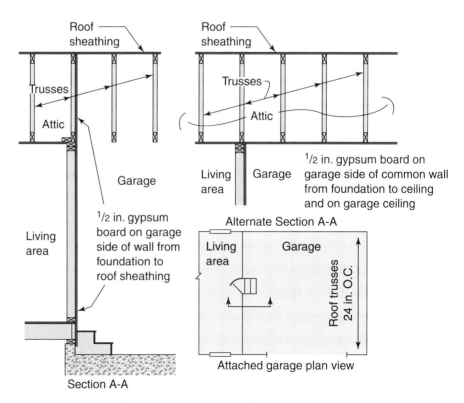

Attached garage separation from dwelling

R302.6 Dwelling/Garage Fire Separation. The garage shall be separated as required by Table R302.6. Openings in garage walls shall comply with Section R302.5. This provision does not apply to garage walls that are perpendicular to the adjacent dwelling unit wall.

CHANGE SIGNIFICANCE: No technical change to the code is intended with this revision to place the garage/dwelling separation requirements in an easier to read table format. For attached garages and detached garages less than 3 feet from the dwelling, the general rule still requires ½-inch gypsum board applied to the garage side of the separation from the dwelling. Where there are habitable rooms above the garage, the separating material must be ⅝-inch Type X gypsum board applied to the ceiling of the garage. The information in the table also intends to maintain the requirement for ½-inch gypsum board applied to the inside surface of the garage walls supporting the floor-ceiling assembly when there are habitable rooms above.

TABLE R302.6   **Dwelling/Garage Separation**

Separation	Material
From the residence and attics	Not less than ½-inch gypsum board or equivalent applied to the garage side
From all habitable rooms above the garage	Not less than ⅝-inch Type X gypsum board or equivalent
Structure(s) supporting floor-ceiling assemblies used for separation required by this section	Not less than ½-inch gypsum board or equivalent
Garages located less than 3 feet (914 mm) from a dwelling unit on the same lot	Not less than ½-inch gypsum board or equivalent applied to the interior side of exterior walls that are within this area

For SI: 1 inch = 25.4 mm, 1 foot = 304.8 mm.

R305.1

Minimum Ceiling Height

CHANGE TYPE: Modification

CHANGE SUMMARY: Ceiling height requirements have been reorganized for clarification, and the exception permitting beam projections below the required ceiling height has been deleted. Bathroom ceiling heights no longer apply to the area above fixtures, provided the fixture is usable. Provisions for lower ceiling heights in portions of basements used for utility and storage have been moved to a separate subsection.

2009 CODE: **R305.1 Minimum Height.** Habitable ~~rooms~~ space, hallways, ~~corridors,~~ bathrooms, toilet rooms, laundry rooms and ~~basements~~ portions of basements containing these spaces shall have a ceiling height of not less than 7 feet (2134 mm). ~~The required height shall be measured from the finish floor to the lowest projection from the ceiling.~~

Exceptions:
1. ~~Beams and girders spaced not less than 4 feet (1219 mm) on center may project not more than 6 inches (152 mm) below the required ceiling height.~~

2. ~~Ceilings in basements without habitable spaces may project to within 6 feet, 8 inches (2032 mm) of the finished floor; and beams, girders, ducts or other obstructions may project to within 6 feet 4 inches (1931 mm) of the finished floor.~~

~~3.~~ **1.** For rooms with sloped ceilings, at least 50 percent of the required floor area of the room must have a ceiling height of at least 7 feet (2134 mm) and no portion of the required floor area may have a ceiling height of less than 5 feet (1524 mm).

~~4.~~ **2.** Bathrooms shall have a minimum ceiling height of 6 feet 8 inches (2036 mm) ~~over the fixture and~~ at the center of the front clearance area for fixtures as shown in Figure R307.1. The ceiling height above fixtures shall be such that the fixture is capable of being used for its intended purpose. A shower or tub equipped with a showerhead shall have a minimum ceiling height of 6 feet 8 inches (2036 mm) above a minimum area 30 inches (762 mm) by 30 inches (762 mm) at the showerhead.

R305.1.1 Basements. Portions of basements that do not contain habitable space, hallways, bathrooms, toilet rooms, and laundry rooms shall have a ceiling height of not less than 6 feet 8 inches (2032 mm).

Exception: Beams, girders, ducts, or other obstructions may project to within 6 feet, 4 inches (1931 mm) of the finished floor.

CHANGE SIGNIFICANCE: The general rule for minimum ceiling height remains at 7 feet. Several changes clarify where the rule applies and what areas qualify for exception. The 7-foot ceiling height now specifically applies to habitable space, as defined in Section R202, hallways, bathrooms, toilet rooms, and laundry rooms. *Corridors* have been removed from the list because the term is not relevant to buildings regulated by the IRC. The instruction on how to measure ceiling height was considered redundant and in conflict with the definition and has been deleted. The exception allowing beams and girders to project below the required ceiling height has also been removed, as there was no discernible purpose for lowering the ceiling height requirement for spaced beams.

The code still allows lower ceiling heights adjacent to fixtures in bathrooms recognizing the practice of occasionally tucking bathrooms under stairs or other spaces with sloped ceilings that do not impede the functionality of the space. However, the provision for maintaining a ceiling height of 6 feet 8 inches at the location of the fixture, such as a lavatory or water closet, was unclear and overly restrictive. The change clarifies that the prescribed ceiling height is measured at the center of the required clearance area in front of the fixture. The area above the fixture may be lower, accommodating a sloped ceiling, provided the fixture is usable.

The lower ceiling height of 6 feet 8 inches also applies to portions of basements that are typically used for utilities or storage—those areas that are not habitable space, hallways, bathrooms, toilet rooms, or laundry rooms. These areas also still allow for beams, girders, ducts, and other objects to project below the required ceiling height, provided they maintain at least a 6 feet 4 inch clearance above the floor. The basement provisions have been moved from an exception to the general rule to a separate subsection stating the requirement in mandatory code language, followed by the exception for typical basement projections. The revised format improves the understanding of the code section.

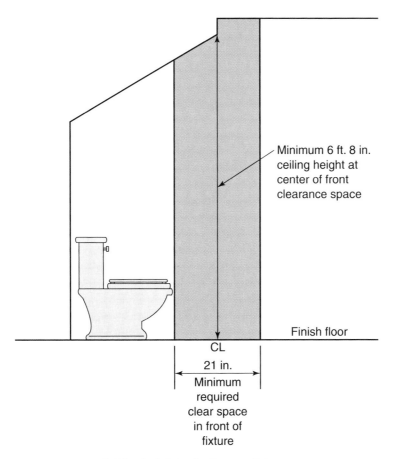

Minimum 6 ft. 8 in. ceiling height at center of front clearance space

Finish floor

CL

21 in.

Minimum required clear space in front of fixture

Ceiling height at bathroom fixtures

R308.1.1 and R308.3

Identification of Glazing and Human Impact Loads

CHANGE TYPE: Modification

CHANGE SUMMARY: The code now recognizes ANSI Z97.1 as an alternative test procedure to CPSC 16 CFR 1201 for safety glazing products not regulated by the federal standard.

2009 CODE: R308.1.1 Identification of Multiple Assemblies. Multipane assemblies having individual panes not exceeding 1 square foot (0.09 m²) in exposed area shall have at least one pane in the assembly identified in accordance with Section R308.1. All other panes in the assembly shall be labeled "CPSC 16 CFR 1201" or "ANSI Z97.1" as appropriate.

R308.3 Human Impact Loads. Individual glazed areas, including glass mirrors in hazardous locations such as those indicated as defined in Section R308.4, shall pass the test requirements of ~~CPSC 16 CFR, Part 1201~~ Section R308.3.1. ~~Glazing shall comply with CPSC 16 CFR, Part 1201 criteria for Category I or Category II as indicated in Table R308.3~~

Exceptions:
1. Louvered windows and jalousies shall comply with Section R308.2.
2. Mirrors and other glass panels mounted or hung on a surface that provides a continuous backing support.
3. Glass unit masonry complying with Section R610.

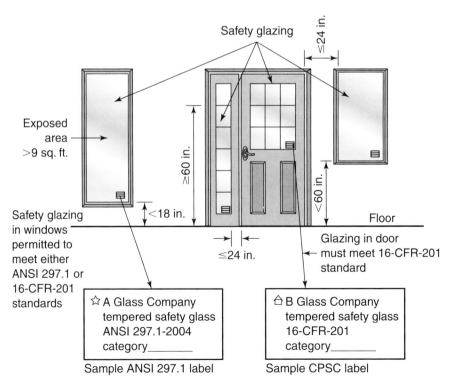

Identification of safety glazing

R308.3.1 Impact Test. Where required by other sections of the code, glazing shall be tested in accordance with CPSC 16 CFR 1201. Glazing shall comply with the test criteria for Category I or II as indicated in Table R308.3.1(1).

> **Exception:** Glazing not in doors or enclosures for hot tubs, whirlpools, saunas, steam rooms, bathtubs and showers shall be permitted to be tested in accordance with ANSI Z97.1. Glazing shall comply with the test criteria for Class A or B as indicated in Table R308.3.1 (2).

CHANGE SIGNIFICANCE: CPSC 16 CFR 1201 *Safety Standard for Architectural Glazing Material* and ANSI Z97.1-2004 *Safety Glazing Materials Used in Buildings—Safety Performance Specifications and Methods of Test*, are standards applicable to determining performance of safety glazing materials used in architectural applications, and both have similar test procedures. The major difference between the two standards is in their scope and function. The U.S. Consumer Products Safety Commission (CPSC) promulgates 16 CFR Part 1201. Unlike the ANSI standard, CPSC 16 CFR 1201 is a federal standard that mandates where and when safety glazing materials must be used in architectural applications and preempts any nonidentical state or local standard. The CPSC requires the installation of safety glazing materials meeting 16 CFR 1201 only in storm doors, combination doors, entrance-exit

R308.1.1 and Table R308.3 continues

TABLE R308.3.1(1)   Minimum Category Classification of Glazing Using CPSC 16 CFR 1201

Exposed Surface Area of One Side of One Lite	Glazing in Storm or Combination Doors (Category Class)	Glazing in Doors (Category Class)	Glazed Panels Regulated by Item 3 of Section R308.4 (Category Class)	Glazed Panels Regulated by Item 2 of Section R308.4 (Category Class)	Glazing in Doors and Enclosures Regulated by Item 5 of Section R308.4 (Category Class)	Sliding Glass Doors Patio Type (Category Class)
9 sq ft or less	I	I	NR	I	II	II
More than 9 sq ft	II	II	II	II	II	II

For SI: 1 square foot = 0.0929 m².
NR means "No Requirement."

TABLE R308.3.1(2)   Minimum Category Classification of Glazing Using ANSI Z97.1

Exposed Surface Area of One Side of One Lite	Glazed Panels Regulated by Item 3 of Section R308.4 (Category class)	Glazed Panels Regulated by Item 2 of Section R308.4 (Category class)	Doors and Enclosures Regulated by Item 5 of Section R308.4[a] (Category class)
9 square feet or less	No requirement	B	A
More than 9 square feet	A	A	A

For SI: 1 square foot = 0.0929 m²
a. Use is permitted only by the Exception to Section R308.3.1.

*R308.1.1 and Table R308.3
continued*

doors, sliding patio doors, closet doors, and shower and tub doors and enclosures. For other locations requiring safety glazing under the IRC, the code now recognizes testing to either standard.

ANSI Z97 Class A glazing materials are comparable to the CPSC Category II glazing materials, passing a 48-inch drop height test, and its Class B glazing materials are comparable to the CPSC's Category I glazing materials, passing the 18-inch drop height test. There are minor differences in the testing methods of the two standards. For impact testing, for example, the CPSC requires only one specimen, where ANSI Z97 requires that four specimens of each nominal thickness and size must be impact-tested. The CPSC standard has no performance tests for plastics or for bent glass. ANSI Z97 has specific tests for both. Thick, heavy annealed glass is likely to pass the CPSC impact tests and is not prohibited in hazardous locations. ANSI Z97.1-2004, on the other hand, does not consider monolithic annealed glass in any thickness to be safety glazing material.

In an editorial change, Exceptions 8 and 10 to Section R308.4 have been relocated as exceptions to Section R308.3. Mirrors and glass panels mounted to a surface providing solid backing and glass block panels are not regulated for safety glazing. Since these items are not exceptions to the general rules in Section R308.4, they more appropriately belong in Section R308.3.

R308.4

Hazardous Locations Requiring Safety Glazing

CHANGE TYPE: Modification

CHANGE SUMMARY: Requirements for safety glazing at hazardous locations subject to human impact have been reorganized in an easy to use format.

2009 CODE: R308.4 Hazardous Locations.   The following shall be considered specific hazardous locations for the purposes of glazing:

1. Glazing in ~~swinging doors except jalousies~~ <u>all fixed and operable panels of swinging, sliding, and bifold doors.</u>

Exceptions:

   1. <u>Glazed openings of a size through which a 3-inch diameter (76 mm) sphere is unable to pass.</u>

   2. <u>Decorative glazing.</u>

~~2. Glazing in fixed and sliding panels of sliding door assemblies and panels in sliding and bifold closet door assemblies.~~

~~3. Glazing in storm doors.~~

~~4. Glazing in all unframed swinging doors.~~

~~6~~<u>2</u>. Glazing in an individual fixed or operable panel adjacent to a door where the nearest vertical edge is within a 24-inch (610 mm) arc of the door in a closed position and whose bottom edge is less than 60 inches (1524 mm) above the floor or walking surface.

R308.4 continues

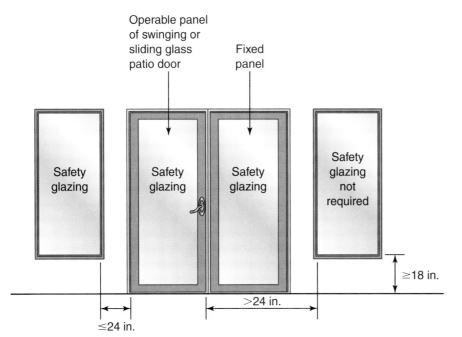

Operable panel of swinging or sliding glass patio door

Fixed panel

Safety glazing

Safety glazing

Safety glazing

Safety glazing not required

≥18 in.

>24 in.

≤24 in.

Glazing adjacent to the fixed panel of patio doors does not require safety glazing

R308.4 continued

Exceptions:
1. Decorative glazing.
2. When there is an intervening wall or other permanent barrier between the door and the glazing.
3. Glazing in walls on the latch side of and perpendicular to the plane of the door in a closed position.
4. Glazing adjacent to a door where access through the door is to a closet or storage area 3 feet (914 mm) or less in depth.
5. Glazing that is adjacent to the fixed panel of patio doors.

~~7~~ **3.** Glazing in an individual fixed or operable panel, ~~other than those locations described in Items 5 and 6 above,~~ that meets all of the following conditions:

~~7.1~~ **3.1.** ~~Exposed~~ The exposed area of an individual pane is larger than 9 square feet (0.836 m^2) and;

~~7.2~~ **3.2.** ~~Bottom~~ The bottom edge of the glazing is less than 18 inches (457 mm) above the floor and;

~~7.3~~ **3.3** ~~Top~~ The top edge of the glazing is more than 36 inches (914 mm) above the floor and;

~~7.4~~ **3.4** One or more walking surfaces are within 36 inches (914 mm), measured horizontally and in a straight line, of the glazing.

Exceptions:
1. Decorative glazing.
2. When a horizontal rail is installed on the accessible side(s) of the glazing 34 to 38 inches (863 to 965 mm) above the walking surface. The rail shall be capable of withstanding a horizontal load of 50 pounds per linear foot (730 N/m) without contacting the glass and be a minimum of 1½ inches (38 mm) in cross sectional height.
3. Outboard panes in insulating glass units and other multiple glazed panels when the bottom edge of the glass is 25 feet (7620 mm) or more above grade, a roof, walking surfaces, or other horizontal [within 45 degrees (0.79 rad) of horizontal] surface adjacent to the glass exterior.

~~8~~ **4.** All glazing in railings regardless of ~~an~~ area or height above a walking surface. Included are structural baluster panels and nonstructural infill panels.

5. Glazing in ~~doors and~~ enclosures for or walls facing hot tubs, whirlpools, saunas, steam rooms, bathtubs and showers. ~~Glazing in any part of a building wall enclosing these compartments~~ where the bottom exposed edge of the glazing is less than 60 inches (1524 mm) measured vertically above any standing or walking surface.

Exception: Glazing that is more than 60 inches (1524 mm), measured horizontally and in a straight line, from the waters edge of a hot tub, whirlpool, or bathtub.

9 6. Glazing in walls and fences ~~enclosing~~ adjacent to indoor and outdoor swimming pools, hot tubs and spas where the bottom edge of the glazing is less than 60 inches (1524 mm) above a walking surface and within 60 inches (1524 mm), measured horizontally and in a straight line, of the water's edge. This shall apply to single glazing and all panes in multiple glazing.

10 7. Glazing adjacent to stairways, landings and ramps within 36 inches (914 mm) horizontally of a walking surface when the exposed surface of the ~~glass~~ glazing is less than 60 inches (1524 mm) above the plane of the adjacent walking surface.

Exceptions:

1. When a rail is installed on the accessible side(s) of the glazing 34 to 38 inches (864 to 965 mm) above the walking surface. The rail shall be capable of withstanding a horizontal load of 50 pounds per linear foot (730 N/m) without contacting the glass and be a minimum of 1½ inches (38 mm) in cross sectional height.

2. The side of the stairway has a guardrail or handrail, including balusters or in-fill panels, complying with Sections R311.7.7 and R312 and the plane of the glazing is more than 18 inches (457 mm) from the railing; or

3. When a solid wall or panel extends from the plane of the adjacent walking surface to 34 inches (864 mm) to 38 inches (965 mm) above the walking surface and the construction at the top of that wall or panel is capable of withstanding the same horizontal load as a guard.

11. 8. Glazing adjacent to stairways within 60 inches (1524 mm) horizontally of the bottom tread of a stairway in any direction when the exposed surface of the ~~glass~~ glazing is less than 60 inches (1524 mm) above the nose of the tread.

Exceptions:

1. The side of the stairway has a guardrail or handrail, including balusters or in-fill panels, complying with R311.7.7 and R312 and the plane of the glass is more than 18 inches (457 mm) from the railing; or

2. When a solid wall or panel extends from the plane of the adjacent walking surface to 34 inches (864 mm) to 38 inches (965 mm) above the walking surface and the construction at the top of that wall or panel is capable of withstanding the same horizontal load as a guard.

Exception: ~~The following products, materials and uses are exempt from the above hazardous locations:~~
~~1. Openings in doors through which a 3-inch (76 mm) sphere is unable to pass.~~

R308.4 continues

R308.4 continued

~~2. Decorative glass in Items 1, 6 or 7.~~

~~3. Glazing in Section R308.4, Item 6, when there is an intervening wall or other permanent barrier between the door and the glazing.~~

~~4. Glazing in Section R308.4, Item 6, in walls perpendicular to the plane of the door in a closed position, other than the wall toward which the door swings when opened, or where access through the door is to a closet or storage area 3 feet (914 mm) or less in depth. Glazing in these applications shall comply with Section R308.4, Item 7.~~

~~5. Glazing in Section R308.4, Items 7 and 10, when a protective bar is installed on the accessible side(s) of the glazing 36 inches ± 2 inches (914 mm ± 51 mm) above the floor. The bar shall be capable of withstanding a horizontal load of 50 pounds per linear foot (730 N/m) without contacting the glass and be a minimum of 11/2 inches (38 mm) in height.~~

~~6. Outboard panes in insulating glass units and other multiple-glazed panels in Section R308.4, Item 7, when the bottom edge of the glass is 25 feet (7620 mm) or more above grade, a roof, walking surfaces, or other horizontal [within 45 degrees (0.79 rad) of horizontal] surface adjacent to the glass exterior.~~

~~7. Louvered windows and jalousies complying with the requirements of Section R308.2.~~

~~8. Mirrors and other glass panels mounted or hung on a surface that provides a continuous backing support.~~

~~9. Safety glazing in Section R308.4, Items 10 and 11, is not required where:~~

~~9.1. The side of a stairway, landing or ramp has a guardrail or handrail, including balusters or in-fill panels, complying with the provisions of Sections 1013 and 1607.7 of the *International Building Code*; and~~

~~9.2. The plane of the glass is more than 18 inches (457 mm) from the railing; or~~

~~9.3. When a solid wall or panel extends from the plane of the adjacent walking surface to 34 inches (863 mm) to 36 inches (914 mm) above the floor and the construction at the top of that wall or panel is capable of withstanding the same horizontal load as the protective bar.~~

~~10. Glass block panels complying with Section R610.~~

CHANGE SIGNIFICANCE: Previously, the code identified 11 hazardous locations requiring safety glazing and, in a separate list, provided 10 exceptions, each of which applied to one or more of the hazardous locations. The arrangement was awkward, making it difficult to quickly identify the applicable requirement. This code change is largely editorial in nature and clarifies the application of the provisions by deleting repetitive or unnecessary language, organizing the material in a logical manner, and moving the exceptions to directly follow the rule to which they apply. The 11 rules have been reduced to 8 by merging the information related to safety glazing in doors.

Item 2, related to glazing adjacent to a door, contains a new Exception 5 intending to clarify that a fixed panel of a sliding door unit is not considered a door in this application. The 24-inch measurement for determining the safety glazing requirement for a sidelight or window is taken from the edge of the active panel only. This provision had apparently been interpreted by some as considering the fixed panel of a sliding door assembly as a "door" and requiring safety glazing for windows that were within 24 inches of the fixed panel.

New text also intends to clarify the requirement for safety glazing adjacent to hot tubs, whirlpools, bathtubs, and showers in Item 5. In addition to enclosures, glazing in walls facing these areas also must be safety glazing if less than 60 inches above the walking surface. This new requirement does not apply to glazing located more than 60 inches horizontally from the water's edge of a hot tub, whirlpool, or bathtub, but there is no such exception for horizontal distance that applies to glazing in walls facing a shower.

The change to item 6 clarifies that safety glazing requirements apply to glazing in fences or walls within 60 inches of the edge of indoor and outdoor swimming pools, hot tubs, and spas, whether or not such fences or walls are part of an enclosure.

R310.1

Emergency Escape and Rescue Openings

CHANGE TYPE: Modification

CHANGE SUMMARY: Habitable attics have been added to the locations requiring an emergency escape and rescue opening.

2009 CODE: R310.1 Emergency Escape and Rescue Required. Basements, habitable attics and every sleeping room shall have at least one operable emergency escape and rescue opening. Such opening shall open directly into a public street, public alley, yard or court. Where basements contain one or more sleeping rooms, emergency egress and rescue openings shall be required in each sleeping room. Where emergency escape and rescue openings are provided they shall have a sill height of not more than 44 inches (1118 mm) above the floor. Where a door opening having a threshold below the adjacent ground elevation serves as an emergency escape and rescue opening and is provided with a bulkhead enclosure, the bulkhead enclosure shall comply with Section R310.3. The net clear opening dimensions required by this section shall be obtained by the normal operation of the emergency escape and rescue opening from the inside. Emergency escape and rescue openings with a finished sill height below the adjacent ground elevation shall be provided with a window well in accordance with Section R310.2. Emergency escape and rescue openings shall open directly into a public way, or to a yard or court that opens to a public way.

CHANGE SIGNIFICANCE: A new defined term in the 2009 IRC, a *habitable attic* is occupiable space between the uppermost floor-ceiling assembly and the roof assembly. The major difference between a habitable attic and other above-ground habitable space is that a habitable attic is not considered a story. For other than basements, the requirement for emergency escape and rescue openings in the 2006 IRC was dependent on the use of the space as a sleeping room. With this change, all habitable attics meeting the definition, whether finished or unfinished and for any use, require an emergency escape and rescue opening.

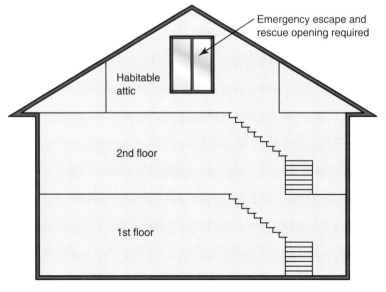

**Emergency escape and
rescue opening for habitable attic**

R311
Means of Egress

CHANGE TYPE: Modification

CHANGE SUMMARY: The means of egress provisions have been reorganized in a systematic order to provide a better understanding of the requirements. New language clarifies that the means of egress in the IRC ends when the occupant reaches grade at the exterior of the building and there are no requirements beyond that point. Net clear opening requirements have replaced the nominal door size for the required egress door to the exterior.

2009 CODE: Section R311 Means of Egress

R311.1 General. ~~Stairways, ramps, exterior egress balconies, hallways and doors shall comply with this section.~~

R311.1 Means of Egress. All dwellings shall be provided with a means of egress as provided in this section. The means of egress shall provide a continuous and unobstructed path of vertical and horizontal egress travel from all portions of the dwelling to the exterior of the dwelling at the required egress door without requiring travel through a garage.

~~R311.4 Doors.~~
~~R311.4.1 Exit Door Required.~~
~~R311.4.2 Door Type and Size.~~
~~R311.4.4 Type of Lock or Latch.~~
R311.2 Egress Door. ~~Not less than~~ At least one ~~exit~~ egress door ~~conforming to this section~~ shall be provided for each dwelling unit. The ~~required exit~~ egress door shall be ~~a~~ side-hinged ~~door, not less than 3 feet (914 mm) in width~~ and shall provide a minimum clear width of 32 inches (813 mm) when measured between the face of the door and the stop, with the door open 90 degrees (1.57 rad). The minimum clear height of the door opening shall not be less than

R311 continues

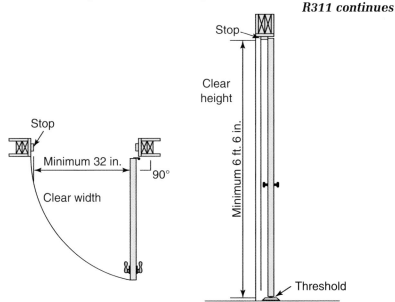

Egress door dimensions

R311 continued   ~~6 feet 8 inches (2032 mm)~~ <u>78 inches (1981 mm) in height measured from the top of the threshold to the bottom of the stop.</u> Other doors shall not be required to comply with these minimum dimensions. ~~All~~ Egress doors shall be readily openable from ~~the side from which egress is to be made~~ <u>inside the dwelling</u> without the use of a key or special knowledge or effort. ~~The required exit door shall provide for direct access from the habitable portions of the dwelling to the exterior without requiring travel through a garage. Access to habitable levels not having an exit in accordance with this section shall be by a ramp in accordance with Section R311.6 or a stairway in accordance with Section R311.5.~~

~~R311.4.3 Landings at Doors.~~

~~There shall be a floor or landing on each side of each exterior door. The floor or landing at the exterior door shall not be more than 1.5 inches (38 mm) lower than the top of the threshold. The landing shall be permitted to have a slope not to exceed 0.25 unit vertical in 12 units horizontal (2-percent).~~

Exceptions:
1. ~~Where a stairway of two or fewer risers is located on the exterior side of a door, other than the required exit door, a landing is not required for the exterior side of the door provided the door, other than an exterior storm or screen door does not swing over the stairway.~~
2. ~~The exterior landing at an exterior doorway shall not be more than 7¾ inches (196 mm) below the top of the threshold, provided the door, other than an exterior storm or screen door does not swing over the landing.~~
3. ~~The height of floors at exterior doors other than the exit door required by Section R311.4.1 shall not be more than 7¾ inches (186 mm) lower than the top of the threshold. The width of each landing shall not be less than the door served. Every landing shall have a minimum dimension of 36 inches (914 mm) measured in the direction of travel.~~

<u>R311.3 Floors and Landings at Exterior Doors.</u>   <u>There shall be a landing or floor on each side of each exterior door. The width of each landing shall not be less than the door served. Every landing shall have a minimum dimension of 36 inches (914 mm) measured in the direction of travel. Exterior landings shall be permitted to have a slope not to exceed 0.25 unit vertical in 12 units horizontal (2-percent).</u>

<u>Exception:</u> <u>Exterior balconies less than 60 square feet (5.6 m^2) and only accessible from a door are permitted to have a landing less than 36 inches (914 mm) measured in the direction of travel.</u>

<u>R311.3.1 Floor Elevations at the Required Egress Door.</u>
<u>Landings or floors at the required egress door shall not be more than 1½ inches (38 mm) lower than the top of the threshold.</u>

Exception: The exterior landing or floor shall not be more than 7¾ inches (196 mm) below the top of the threshold provided the door does not swing over the landing or floor.

When exterior landings or floors serving the required egress door are not at grade, they shall be provided with access to grade by means of a ramp in accordance with Section R311.8 or a stairway in accordance with Section R311.7.

R311.3.2 Floor Elevations for Other Exterior Doors. Doors other than the required egress door shall be provided with landings or floors not more than 7¾ inches (196 mm) below the top of the threshold.

Exception: A landing is not required where a stairway of two or fewer risers is located on the exterior side of the door, provided the door does not swing over the stairway.

R311.3.3 Storm and Screen Doors. Storm and screen doors shall be permitted to swing over all exterior stairs and landings.

R311.4 Vertical Egress. Egress from habitable levels, including habitable attics and basements not provided with an egress door in accordance with Section R311.2 shall be by a ramp in accordance with Section R311.8 or a stairway in accordance with Section R311.7.

R311.2 R311.5 Construction.
R311.2.1 R311.5.1 Attachment. ~~Required exterior egress balconies, exterior exit stairways and similar means of egress components~~ Exterior landings, decks, balconies, stairs and similar facilities shall be positively anchored to the primary structure to resist both vertical and lateral forces or shall be designed to be self-supporting. ~~Such~~ Attachment shall not be accomplished by use of toenails or nails subject to withdrawal.

R311.2.2 Under Stair Protection. ~~Enclosed accessible space under stairs shall have walls, under stair surface and any soffits protected on the enclosed side with 1/2-inch (12.7 mm) gypsum board.~~

R311.3 R311.6 Hallways. The minimum width of a hallway shall be not less than 3 feet (914 mm).

CHANGE SIGNIFICANCE: Previously, Section R311 began by identifying the elements of the means of egress system. That language has been replaced with a charging statement requiring a means of egress for all dwellings followed by a performance-based definition requiring a continuous and unobstructed path to the exterior from any location in the dwelling. The revision of the entire section organizes the provisions in a user-friendly format but intends only minor technical changes. *Habitable attics* are now defined in the IRC and require a stair or ramp meeting the egress provisions of Section R311.

R311 continues

R311 continued

The means of egress for dwellings ends when the occupant exits the building and reaches grade. This point is clarified by the definition of *means of egress* in Section R311.1 stipulating a travel path to the exterior of the dwelling at the required egress door. In addition, Section R311.3.1 requires the exterior landing of the required egress door to be at grade or to provide access to grade by means of a ramp or stairway.

The word "egress" replaces the word "exit" for the requirement of at least one egress door in Section R311.2. Section R311.1 describes the required egress door as an exterior door for exiting the building to reach the outside. New language in Section R311.3.1 clarifies that the landing at the required egress door must be at grade, or the code requires access to grade with a ramp or stairs. Though the means of egress must provide a continuous path to the outside, other doors are no longer specifically identified as egress doors (2006 IRC Section R311.1). Where the code previously required that all egress doors be readily openable in the direction of egress, now only the one required egress door must be readily openable from the inside of the building. The restriction on the lock or latch to be opened without the use of a key or special knowledge or effort also now only applies to the one required egress door.

Previously, a door with a nominal size of 3 feet 0 inches by 6 feet 8 inches, a common industry standard, satisfied the size requirements for the one required exit door. The code now specifies the required net clear opening dimensions and the method for measuring when the door is opened to the 90 degree position. The minimum net opening dimensions are now consistent with the door requirements for means of egress and accessibility for persons with disabilities in the IBC.

A new exception to the landing requirement at exterior doors clarifies that the intent of the 36-inch dimension in the direction of travel is to provide a safe transition when exiting the building. It is not the intent of the code to prohibit guards that are applied directly to the exterior side of a door that is not the required egress door or to prohibit balconies with smaller dimensions. The 2009 IRC specifically excludes small viewing or decorative balconies from this landing dimension requirement when the balcony can only be accessed through a door. Also, the code now specifically recognizes that exterior landings, decks, and other such elements are permitted to be self-supporting or be securely attached to the dwelling structure.

The requirement for protecting enclosed spaces under stairs with ½-inch gypsum board has been moved from Section R311.2.2 to the fire-resistant construction provisions in Section R302.7.

CHANGE TYPE: Modification

R311.7.2
Stairway Headroom

CHANGE SUMMARY: The added language clarifies that the minimum headroom is measured above the usable area of the treads in an open stairway and does not apply to the ends of treads where they project under the edge of the floor opening above.

2009 CODE: R311.5.2 R311.7.2 Headroom. The minimum headroom in all parts of the stairway shall not be less than 6 feet 8 inches (2036 mm) measured vertically from the sloped ~~plane~~ line adjoining the tread nosing or from the floor surface of the landing or platform on that portion of the stairway.

> **Exception:** Where the nosings of treads at the side of a flight extend under the edge of a floor opening through which the stair passes, the floor opening shall be allowed to project horizontally into the required headroom a maximum of 4¾ inches (121 mm).

CHANGE SIGNIFICANCE: The code now more clearly states the intent that minimum stair headroom height is required above only the area where a person normally walks on the stair. The revision permits the edge of the floor opening above to be approximately in line with the inside edge of the guard on an open side of a stair below. This permits stair openings and support walls to be positioned in line in the vertical plane without creating any hazard.

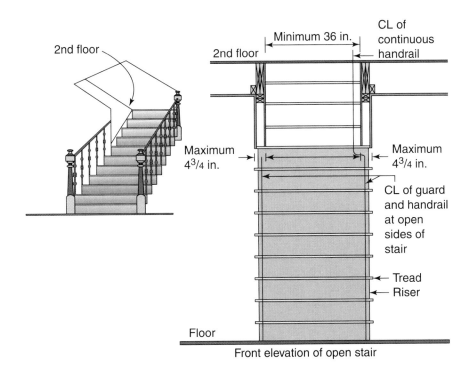

Front elevation of open stair

Stair headroom at open stair

R311.7.3 and R311.7.4

Stair Treads and Risers

CHANGE TYPE: Modification

CHANGE SUMMARY: New provisions defining the walk line intend to clarify the tread depth requirement for winders. Measurement of stair treads and risers exclusive of carpet will result in more consistent application of the code. A new exception to uniform tread depth provides that winders are permitted for a turn in a stairway of otherwise rectangular treads.

2009 CODE: <u>**R311.7.3 Walk Line.** The walk line across winder treads shall be concentric to the curved direction of travel through the turn and located 12 inches (305 mm) from the side where the winders are narrower. The 12 inch (305 mm) dimension shall be measured from the widest point of the clear stair width at the walking surface of the winder. If winders are adjacent within the flight, the point of the widest clear stair width of the adjacent winders shall be used.</u>

~~**R311.5.3**~~ <u>**R311.7.4**</u> **Stair Treads and Risers.** <u>Stair treads and risers shall meet the requirements of this section. For the purposes of this section all dimensions and dimensioned surfaces shall be exclusive of carpets, rugs, or runners.</u>

~~**R311.5.3.1**~~ <u>**R311.7.4.1**</u> **Riser Height.** (No change to text.)

~~**R311.5.3.2**~~ <u>**R311.7.4.2**</u> **Tread Depth.** The minimum tread depth shall be 10 inches (254 mm). The tread depth shall be measured horizontally between the vertical planes of the foremost projection of adjacent treads and at a right angle to the tread's leading edge. The greatest tread depth within any flight of stairs shall not exceed the smallest by more than $\frac{3}{8}$ inch (9.5 mm).

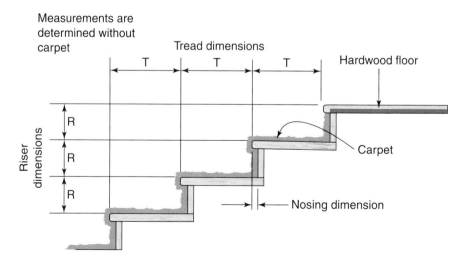

Stair tread and riser measurements

<u>Consistently shaped winders at the walk line shall be allowed within the same flight of stairs as rectangular treads and do not have to be within ³⁄₈ inch (9.5 mm) of the rectangular tread depth.</u> Winder treads shall have a minimum tread depth of 10 inches (254 mm) measured ~~as above at a point 12 inches (305 mm) from the side where the treads are narrower~~ <u>between the vertical planes of the foremost projection of adjacent treads at the intersections with the walk line.</u> Winder treads shall have a minimum tread depth of 6 inches (152 mm) at any point <u>within the clear width of the stair.</u> Within any flight of stairs, the largest winder tread depth at the ~~12- inch (305 mm)~~ walk line shall not exceed the smallest by more than ³⁄₈ inch (9.5 mm).

~~R311.5.3.3~~ R311.7.4.3 Profile.   The radius of curvature at the ~~leading edge of the tread~~ <u>nosing</u> shall be no greater than $^{9}/_{16}$ inch (14 mm). A nosing not less than ¾ inch (19 mm) but not more than 1¼ inch (32 mm) shall be provided on stairways with solid risers. The greatest nosing projection shall not exceed the smallest nosing projection by more than ³⁄₈ inch (9.5 mm) between two stories, including the nosing at the level of floors and landings. Beveling of nosings shall not exceed ½ inch (12.7 mm). Risers shall be vertical or sloped under the tread above from the underside of the ~~leading edge of the tread~~ <u>nosing</u> above at an angle not more than 30 degrees (0.51 rad) from the vertical. Open risers are permitted, provided that the opening between treads does not permit the passage of a 4-inch diameter (102 mm) sphere.

Exceptions:

1. A nosing is not required where the tread depth is a minimum of 11 inches (279 mm).
2. The opening between adjacent treads is not limited on stairs with a total rise of 30 inches (762 mm) or less.

CHANGE SIGNIFICANCE: For winder treads, the addition of Section R311.7.3 intends to standardize the walk line location for determining tread depth and provide for consistent application of the winder provisions. Measuring the tread depth at the intersection with the walk line results in winders that are uniform in depth at the most common path of travel.

Carpet is not regulated by the code but is commonly considered in measuring stair riser height. Because carpet and padding vary in thickness and are compressible, accurate measurement is difficult. In addition, carpet may not be installed at the time of final inspection or initial occupancy, but may be installed or replaced at a later date. Such is common for basement stairs where carpeting is installed at some future date when the basement is finished. It is not the intent of the code to regulate future carpet installations. The IRC now provides for measurement of stair risers before carpet is installed. The objective and accurate measurement will result in more uniform application of the stair riser requirements.

R311.7.3 and R311.7.4 continues

R311.7.3 and R311.7.4 continued

Though it is common practice to use winders to make a turn in a stairway with otherwise rectangular treads without the use of landings, the code has not specifically addressed this installation. Some code users have interpreted the tread uniformity provisions to prohibit this practice. The new exception clarifies that the two different tread types are permitted to occur in the same flight of stairs. While each type of stair tread must meet the respective uniformity requirements when compared with a tread of the same type, the winder tread depth at the walk line is not required to match the tread depth of rectangular steps (plus or minus the $\frac{3}{8}$-inch tolerance) when in the same flight of stairs. Further modification to Section R311.7.4.2 reflects the new determination for walk line in Section R311.7.3 and deletes redundant or conflicting text.

R311.7.7
Handrails

CHANGE TYPE: Modification

CHANGE SUMMARY: Transition fittings are now permitted to exceed the maximum handrail height of 38 inches. An editorial change clarifies that Type I handrails must have rounded edges consistent with the description of Type II handrails.

2009 CODE: ~~R311.5.6~~ <u>R311.7.7</u> Handrails.   Handrails shall be provided on at least one side of each continuous run of treads or flight with four or more risers.

~~R311.5.6.1~~ <u>R311.7.7.1</u> Height.   Handrail height, measured vertically from the sloped plane adjoining the tread nosing, or finish surface of ramp slope, shall be not less than 34 inches (864 mm) and not more than 38 inches (965 mm).

<u>Exceptions:</u>
<u>**1.** The use of a volute, turnout, or starting easing shall be allowed over the lowest tread.</u>

<u>**2.** When handrail fittings or bendings are used to provide continuous transition between flights, the transition from handrail to guardrail, or used at the start of a flight, the handrail height at the fittings or bendings shall be permitted to exceed the maximum height.</u>

~~R311.5.6.2~~ <u>R311.7.7.2</u> Continuity.   (No change to text.)

~~R311.5.6.3~~ <u>R311.7.7.3</u> Handrail Grip Size.   All required handrails shall be of one of the following types or provide equivalent graspability.
   1. Type I. Handrails with a circular cross section shall have an outside diameter of at least 1¼ inches (32 mm) and not greater than 2 inches (51 mm). If the handrail is not circular, it shall have a perimeter dimension of at least 4 inches (102 mm) and not greater than 6¼ inches (160 mm) with a maximum cross section dimension of 2¼ inches (57 mm). <u>Edges shall have a minimum radius of 0.01 inch (0.25 mm).</u>
   2. Type II. (No change to text.)

CHANGE SIGNIFICANCE: In the continuity provisions for handrails, a starting easing is permitted over the lowest tread. Rather than matching the slope of the stair, this type of fitting is typically level, causing it to exceed the handrail height at the lowest tread location. Similarly, some terminal fittings installed at the top of a stairway or those providing a transition between sections of handrail rise higher than 38 inches above the plane of the tread nosings. The new exceptions recognize that transition fittings used for continuity or terminations of handrails are permitted to exceed the maximum handrail height of 38 inches.

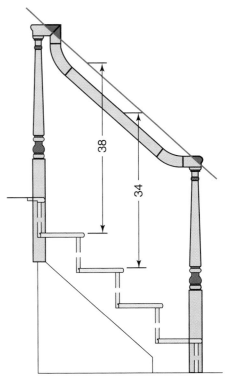

Handrail height at transition fittings

R312
Guards

CHANGE TYPE: Modification

CHANGE SUMMARY: When determining where a guard is required, the vertical distance from the walking surface to the grade or floor below is measured to the lowest point within 36 inches horizontally from the edge of the open-sided walking surface. In determining the minimum height of a guard, fixed seating is considered the same as a

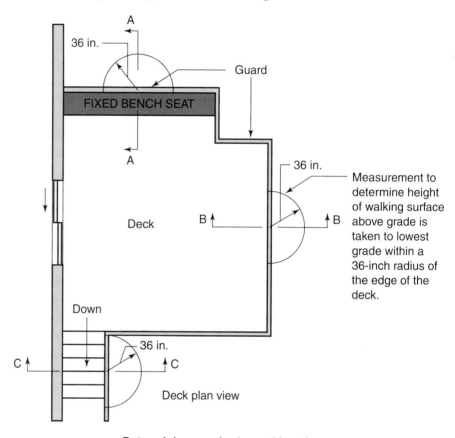

Determining required guard locations

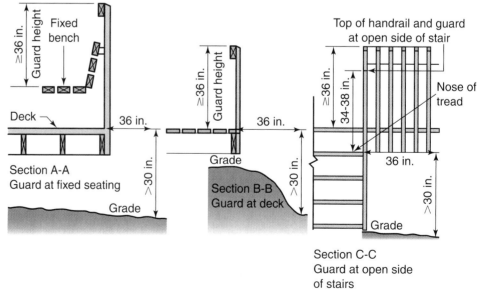

Determining required guard locations

walking surface, and the height is measured from the seat to the top of the guard.

2009 CODE: Section R312 Guards

R312.1 ~~Guards~~ Where required. ~~Porches, balconies, ramps or raised floor surfaces located more than 30 inches (762 mm) above the floor or grade below shall have guards not less than 36 inches (914 mm) in height. Open sides of stairs with a total rise of more than 30 inches (762 mm) above the floor or grade below shall have guards not less than 34 inches (864 mm) in height measured vertically from the nosing of the treads.~~ Guards shall be located along open-sided walking surfaces, including stairs, ramps and landings, that are located more than 30 inches (762 mm) measured vertically to the floor or grade below at any point within 36 inches (914 mm) horizontally to the edge of the open side. Insect screening shall not be considered as a guard.

~~Porches and decks which are enclosed with insect screening shall be equipped with guards where the walking surface is located more than 30 inches (762 mm) above the floor or grade below.~~

R312.2 Height. Required guards at open-sided walking surfaces, including stairs, porches, balconies or landings, shall be not less than 36 inches (914 mm) high measured vertically above the adjacent walking surface, adjacent fixed seating or the line connecting the leading edges of the treads.

Exceptions:
1. Guards on the open sides of stairs shall have a height not less than 34 inches (864 mm) measured vertically from a line connecting the leading edges of the treads.
2. Where the top of the guard also serves as a handrail on the open sides of stairs, the top of the guard shall not be not less than 34 inches (864 mm) and not more than 38 inches (965 mm) measured vertically from a line connecting the leading edges of the treads.

~~R312.2~~ R312.3 Guard Opening Limitations. Required guards ~~on open sides of stairways, raised floor areas, balconies and porches~~ shall **not** have openings from the walking surface to the required guard height ~~intermediate rails or ornamental closures~~ which ~~do not~~ allow passage of a sphere 4 inches (102 mm) ~~or more~~ in diameter.

Exceptions:
1. The triangular openings at the open side of a stair, formed by the riser, tread and bottom rail of a guard, ~~at the open side of a stairway are permitted to be of such a size that a sphere 6 inches cannot pass through.~~ shall not allow passage of a sphere 6 inches (153 mm) in diameter.
2. ~~Openings for required guards on the open sides of stair treads stairs shall not allow passage of a sphere 4 3/8 inches or more in diameter to pass through~~ Guards on the open sides of stairs

R312 continues

R312 continued

shall not have openings which allow passage of a sphere 4⅜ inches (111 mm) in diameter.

CHANGE SIGNIFICANCE: The provisions for guards have been reorganized into three separate sections—required locations, height, and opening limitations—and the technical provisions revised for clarification and consistency in application. The first section now only determines where guards are required and gives an objective means for measuring the height of the walking surface above the grade below. Previously, in the case of a deck, this measurement was typically taken to the grade directly below the edge of the decking surface. However, a sloped site or sudden drop-off adjacent to a deck or porch caused concern that such a measurement did not accurately reflect the level of hazard. The added language requires the height of the walking surface above grade to be measured from the lowest point within 3 feet horizontally from the edge of the deck, porch, or other element. On a sloped site, the measurement to the lowest point may necessitate a guard where one was not specifically required under the 2006 IRC.

This section also removes the incomplete list of example elements requiring guards (porches, balconies, and raised floor surfaces) and applies the generic description of all open-sided walking surfaces, including stairs, ramps and landings, that are within the scope of the IRC—that is, walking surfaces that are part of dwellings or accessory buildings—when they exceed the criteria for height above the grade or floor below. The IRC does not regulate sidewalks, driveways, yards, or typical residential landscaping elements for guard requirements.

The guard location requirements no longer specifically identify a screened porch as requiring a guard, but stipulate that insect screening is not, in itself, considered a guard. The revised language still permits some discretion by the building official in determining the performance of screen and other materials as infill components able to withstand a horizontally applied normal load of 50 pounds applied to a one square foot area.

Minimum guard height requirements have been placed in a separate section following the location provisions. The change clarifies the intent of the code that guard dimensions apply only to required guards, not to optional barriers that are installed where guards are not required by the code. For the most part, the technical provisions for the height of guards and the exceptions for guards at the side of open stairs remain the same. The significant change is that fixed seating adjacent to a guard is now considered in determining guard height, and the vertical measurement is taken from the surface of the seat to the top of the guard. The new requirement is in response to concerns of children climbing or playing on the seat with the potential to fall over a guard, the height of which was previously measured from the floor. There is no realistic way to regulate movable furniture or decorative elements and this new provision only applies to benches or seating surfaces that are secured in place.

Editorial changes to the third section on guard opening dimensions provide clarification and remove unnecessary language but do not change the technical provisions.

R313
Automatic Fire Sprinkler Systems

CHANGE TYPE: Addition

CHANGE SUMMARY: An automatic fire sprinkler system installed in accordance with IRC Section P2904 or NFPA 13D is now required for one- and two-family dwellings and townhouses. The townhouse requirements are effective upon adoption of the 2009 IRC, while the one- and two-family dwelling provisions are not effective until January 1, 2011.

2009 CODE:
Section R313
Automatic Fire Sprinkler Systems
R313.1 Townhouse Automatic Fire Sprinkler Systems. An automatic residential fire sprinkler system shall be installed in townhouses.

> **Exception:** An automatic residential sprinkler system shall not be required when additions or alterations are made to existing townhouses that do not have an automatic residential fire sprinkler system installed.

R313.1.1 Design and Installation. Automatic residential fire sprinkler systems for townhouses shall be designed and installed in accordance with Section P2904.

R313.2 One- and Two-Family Dwellings Automatic Fire Sprinkler Systems. Effective January 1, 2011, an automatic residential fire sprinkler system shall be installed in one- and two-family dwellings.

Courtesy of Uponor Inc.

R313 continues

R313 continued

Exception: An automatic residential fire sprinkler system shall not be required for additions or alterations to existing buildings that are not already provided with an automatic residential fire sprinkler system.

R313.2.1 Design and installation. Automatic residential fire sprinkler systems shall be designed and installed in accordance with Section P2904 or NFPA 13D.

~~APPENDIX P~~
~~Fire Sprinkler System~~
~~The provisions contained in this appendix are not mandatory unless specifically referenced in the adopting ordinance.~~

~~AP101 Fire Sprinklers.~~ ~~An approved automatic fire sprinkler system shall be installed in new one- and two-family dwellings and townhouses in accordance with Section 903.3.1 of the~~ *~~International Building Code.~~*

CHAPTER 44: REFERENCED STANDARDS
NFPA 13D-07 Installation of Sprinkler Systems in One- and Two-Family Dwellings and Manufactured Homes

CHANGE SIGNIFICANCE: An automatic sprinkler system conforming to IRC Section P2904 or NFPA 13D, *Installation of Sprinkler Systems in One- and Two-Family Dwellings and Manufactured Homes*, is now required fire protection in all new one- and two-family dwellings and townhouses. The fire sprinkler provisions for one- and two-family dwellings take effect January 1, 2011. In townhouses, requirements for a residential automatic sprinkler system take effect upon adoption of the 2009 IRC. A dwelling automatic sprinkler system aids in the detection and control of fires in residential occupancies. When installed in accordance with either of the referenced standards, the automatic sprinkler system is expected to prevent total fire involvement (flashover) in the room of fire origin if it is sprinklered. A properly installed and maintained automatic sprinkler system complying with IRC Section P2904 or NFPA 13D improves the likelihood of occupants escaping or being evacuated.

A dwelling fire sprinkler system requires less water when compared with NFPA 13 and 13R systems. Section P2904.5.2 requires a minimum water discharge duration of 10 minutes, compared with 30 minutes for an NFPA 13R system and even higher duration values in NFPA 13. The minimum sprinkler discharge density for an NFPA 13D automatic sprinkler system may be satisfied by connection to a domestic water supply, a water well, an elevated storage tank, an approved pressure tank, or a stored water source with an automatically operated pump. Any combination of water supply systems is allowed to meet the required dwelling fire sprinkler system capacity.

Compared with an NFPA 13 automatic sprinkler system, a dwelling fire sprinkler system does not require automatic sprinkler protection throughout a one- and two-family dwelling or townhouse. Sprinklers are not required in areas that have been statistically shown through fire incident loss data to not significantly contribute to injuries or death. Section P2904.1.1 omits sprinklers in closets and pantries with an area less than 24 square feet constructed of gypsum board walls and ceilings, and bathrooms with an area of 55 square feet or less. Sprinklers are not required in open attached porches, garages, attics, crawl spaces, and concealed spaces not intended or used for living purposes. In attics housing fuel-fired appliances, Section 2904.1.1 requires a single sprinkler above the equipment but does not require that the sprinkler protection be provided throughout the entire attic area. A dwelling fire sprinkler system does not require a fire department connection.

IRC Section P2904 provides a simple, prescriptive approach to the design of dwelling fire sprinkler systems. These provisions will allow a contractor or home builder to design and install a residential sprinkler system without referencing NFPA 13D. The requirements in Section P2904 are consistent with the requirements in NFPA 13D but have been simplified. Design professionals still have the option of using NFPA 13D, which allows for engineered design options and other piping configurations.

R314

Smoke Alarms

CHANGE TYPE: Clarification

CHANGE SUMMARY: Reorganization of the smoke alarm provisions places all of the power requirements in one section and separates the alternative household fire alarm systems from the smoke alarm section. New text clarifies the maintenance and supervision requirements for household fire alarm systems. An exception provides that a household fire alarm system, where installed in addition to smoke alarms, is not required to provide the same level of protection. Habitable attics have been added to the list of locations requiring smoke alarms. Minor plumbing and mechanical work is now permitted without triggering the smoke alarm provisions in existing dwellings.

2009 CODE: **Section R313 R314 Smoke Alarms**
R313.1 R314.1 Smoke Detection and Notification. All smoke alarms shall be listed in accordance with UL 217 and installed in accordance with the provisions of this code and the household fire warning equipment provisions of NFPA 72.

R314.2 Smoke Detection Systems. Household fire alarm systems installed in accordance with NFPA 72 that include smoke alarms, or a combination of smoke detector and audible notification device installed as required by this section for smoke alarms, shall be permitted. The household fire alarm system shall provide the same level of smoke detection and alarm as required by this section for smoke alarms in the event the fire alarm panel is removed or the system is not connected to a central station. Where a household fire warning system is installed using a combination of smoke detector and audible notification device(s), it shall become a permanent fixture of the occupancy

R314 continued

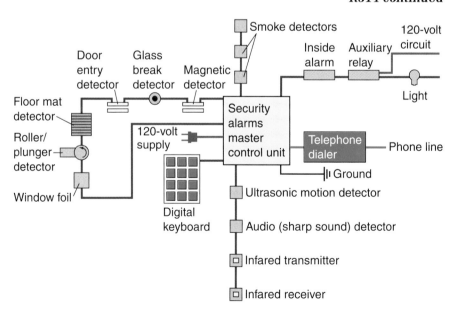

Household fire alarm system

and owned by the homeowner. The system shall be monitored by an approved supervising station and be maintained in accordance with NFPA 72.

> **Exception:** Where smoke alarms are provided meeting the requirements of Section R314.4.

R313.2 R314.3 Location.   Smoke alarms shall be installed in the following locations:

1. In each sleeping room.

2. Outside each separate sleeping area in the immediate vicinity of the bedrooms.

3. On each additional story of the dwelling, including basements and habitable attics, but not including crawl spaces and uninhabitable attics. In dwellings or dwelling units with split levels and without an intervening door between the adjacent levels, a smoke alarm installed on the upper level shall suffice for the adjacent lower level provided that the lower level is less than one full story below the upper level.

When more than one smoke alarm is required to be installed within an individual dwelling unit, the alarm devices shall be interconnected in such a manner that the actuation of one alarm will activate all of the alarms in the individual unit.

R313.2.1 R314.3.1 Alterations, Repairs and Additions.   When alterations, repairs or additions requiring a permit occur, or when one or more sleeping rooms are added or created in existing dwellings, the individual dwelling unit shall be equipped with smoke alarms located as required for new dwellings; the smoke alarms shall be interconnected and hard wired.

> **Exceptions:**
>
> 1. Inter connection and hard-wiring of smoke alarms in existing areas shall not be required where the alterations or repairs do not result in the removal of interior wall or ceiling finishes exposing the structure, unless there is an attic, crawl space or basement available which could provide access for hard wiring and interconnection without the removal of interior finishes.
>
> 2. 1. Work involving the exterior surfaces of dwellings, such as the replacement of roofing or siding, or the addition or replacement of windows or doors, or the addition of a porch or deck, are exempt from the requirements of this section.
>
> 2. Installation, alteration or repairs of plumbing or mechanical systems are exempt from the requirements of this section.

R313.3 R314.4 Power Source.   In new construction, the required Smoke alarms shall receive their primary power from the building wiring when such wiring is served from a commercial source, and when primary power is interrupted, shall receive power from a battery. Wiring shall be permanent and without a disconnecting switch

R314 continues

R314 continued other than those required for overcurrent protection. Smoke alarms shall be interconnected. ~~Smoke alarms shall be permitted to be battery operated when installed in buildings without commercial power or in buildings that undergo alterations, repairs or additions regulated by Section R313.2.1.~~

Exceptions:

1. Smoke alarms shall be permitted to be battery operated when installed in buildings without commercial power.

2. Interconnection and hard-wiring of smoke alarms in existing areas shall not be required where the alterations or repairs do not result in the removal of interior wall or ceiling finishes exposing the structure, unless there is an attic, crawl space or basement available which could provide access for hard-wiring and interconnection without the removal of interior finishes.

CHANGE SIGNIFICANCE: A smoke alarm is a self-contained device that provides both smoke detection and an alarm sounding appliance. Smoke alarms must be listed as conforming to UL 217, *Single and Multiple Station Smoke Alarms*. The IRC permits a household fire alarm system, which typically has separate devices for smoke detection and alarm annunciation, as an alternative to smoke alarms. This distinction is clarified by placing the applicable provisions in separate sections. The requirement in the 2006 IRC for the household fire alarm system to operate if the panel was removed has been deleted and language added to clarify these provisions. Such a system cannot function if the fire alarm panel is removed. The added language ensures system reliability by requiring the system to be owned by the occupant and to be electronically monitored and maintained in accordance with the referenced standard. In situations where both smoke alarms and a household fire alarm system are installed, the household fire alarm system is not required to provide the same level of smoke detection and alarm notification as the smoke alarms.

In general, when construction work involving an existing dwelling occurs, the building must be brought into compliance with the smoke alarm provisions as for a new dwelling. Such provisions do not apply in the case of minor work that does not require a permit, for exterior renovations, or for the addition of decks or porches. A new exception permits plumbing and mechanical work on an existing dwelling without bringing the building into compliance with the smoke alarm provisions.

Habitable attics, a new defined term in the code, has been added to the list of locations requiring a smoke alarm. Placing all smoke alarm power requirements in one section eliminates redundant language and clarifies the two instances where smoke alarms powered by only batteries are permitted.

R315

Carbon Monoxide Alarms

CHANGE TYPE: Addition

CHANGE SUMMARY: The 2009 IRC requires carbon monoxide alarms in new dwellings and in existing dwellings when work requiring a permit takes place. The carbon monoxide alarms must be installed in the immediate vicinity of sleeping areas.

2009 CODE: **Section R315 Carbon Monoxide Alarms**

R315.1 Carbon Monoxide Alarms. For new construction, an approved carbon monoxide alarm shall be installed outside of each separate sleeping area in the immediate vicinity of the bedrooms in dwelling units within which fuel-fired appliances are installed and in dwelling units that have attached garages.

R315.2 Where Required in Existing Dwellings. Where work requiring a permit occurs in existing dwellings that have attached garages or in existing dwellings within which fuel-fired appliances exist, carbon monoxide alarms shall be provided in accordance with Section R315.1.

R315.3 Alarm Requirements. Single station carbon monoxide alarms shall be listed as complying with UL 2034 and shall be installed in accordance with this code and the manufacturer's installation instructions.

CHANGE SIGNIFICANCE: Carbon monoxide alarms are now required in new dwelling units constructed under the 2009 IRC. Because the source of unsafe levels of carbon monoxide in the home is typically from faulty operation of a fuel-fired furnace or water heater, or from the exhaust of an automobile, this new requirement applies only to homes containing fuel-fired appliances or having an attached garage. Carbon

R315 continues

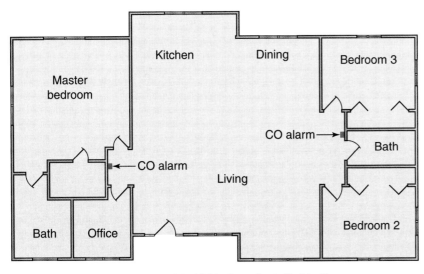

**Carbon monoxide (CO) alarm installed in the
immediate vicinity of each sleeping area**

R315 continued   monoxide accumulates in the body over time relative to its concentration in the air. Accordingly, carbon monoxide detectors sound an alarm based on the concentration of carbon monoxide and the amount of time that certain levels are detected, simulating an accumulation of the toxic gas in the body. High levels of carbon monoxide will trigger an alarm within a short time, while lower levels must be present over a longer time period for the alarm to sound. This design prevents false-positive alarms. This change in the code recognizes the improved reliability of carbon monoxide alarms and the referenced standard, UL 2034, *Single and Multiple Station Carbon Monoxide Alarms*, and intends to reduce accidental deaths from carbon monoxide poisoning.

Under the new provisions, carbon monoxide alarms are also required in existing dwelling units. Similar to the smoke alarm provisions, this requirement for installation of carbon monoxide alarms is triggered by construction work on the existing dwelling where such work requires a permit. Unlike the smoke alarm requirements, there is no exception for exterior work or the addition of decks or porches. Roofing, siding, window replacement, and other exterior work requiring a permit will require the installation of carbon monoxide alarms.

Because carbon monoxide poisoning deaths often occur when the occupant is sleeping, the IRC requires carbon monoxide alarms to be located in the areas outside of and adjacent to bedrooms.

CHANGE TYPE: Modification

CHANGE SUMMARY: Protection from decay is now required for wood siding, sheathing, and wall framing less than 2 inches above a concrete slab exposed to weather.

2009 CODE: Section ~~R319~~ R317 Protection of Wood and Wood Based Products Against Decay

~~R319.1~~ R317.1 Location Required.   Protection of wood and wood based products from decay shall be provided in the following locations by the use of naturally durable wood or wood that is preservative treated in accordance with AWPA U1 for the species, product, preservative and end use. Preservatives shall be listed in Section 4 of AWPA U1.

1. Wood joists or the bottom of a wood structural floor when closer than 18 inches (457 mm) or wood girders when closer than 12 inches (305 mm) to the exposed ground in crawl spaces or unexcavated area located within the periphery of the building foundation.

2. All wood framing members that rest on concrete or masonry exterior foundation walls and are less than 8 inches (203 mm) from the exposed ground.

3. Sills and sleepers on a concrete or masonry slab that is in direct contact with the ground unless separated from such slab by an impervious moisture barrier.

4. The ends of wood girders entering exterior masonry or concrete walls having clearances of less than ½ inch (12.7 mm) on tops, sides and ends.

R317.1 continues

R317.1

Locations for Protection Against Decay

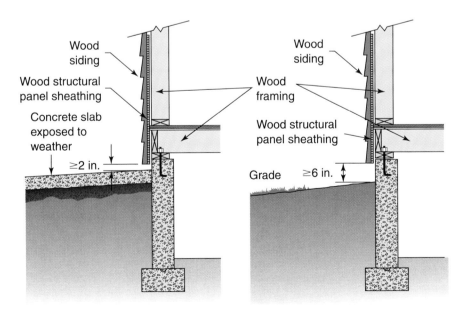

Protection against decay

R317.1 continued

5. Wood siding, sheathing and wall framing on the exterior of a building having a clearance of less than 6 inches (152 mm) from the ground <u>or less than 2 inches (51 mm) measured vertically from concrete steps, porch slabs, patio slabs, and similar horizontal surfaces exposed to the weather.</u>

6. Wood structural members supporting moisture-permeable floors or roofs that are exposed to the weather, such as concrete or masonry slabs, unless separated from such floors or roofs by an impervious moisture barrier.

7. Wood furring strips or other wood framing members attached directly to the interior of exterior masonry walls or concrete walls below grade except where an approved vapor retarder is applied between the wall and the furring strips or framing members.

CHANGE SIGNIFICANCE: Where less than 6 inches above the ground, exterior wood siding, sheathing, and wall framing require protection from decay using naturally durable or preservative-treated wood. The new language requiring a 2-inch clearance above an exterior concrete slab or similar surface is consistent in that a typical concrete slab has a nominal thickness of 4 inches. The clearances are necessary to prevent the wood from absorbing water, resulting in decay, and to provide sufficient space to check for termite activity.

CHANGE TYPE: Modification

CHANGE SUMMARY: The fastener requirements have been expanded to include fasteners and connectors in contact with preservative-treated and fire-retardant-treated wood. New subsections distinguish between fire-retardant-treated wood (FRTW) in exterior and interior applications, deferring to the manufacturer's recommendations for fasteners in an interior location.

R317.3

Fasteners and Connectors in Contact with Treated Wood

R319.3 R317.3 Fasteners and Connectors in Contact with Preservative-treated and Fire-retardant-treated Wood. Fasteners ~~for~~ and connectors in contact with ~~pressure~~-preservative-treated wood and fire-retardant-treated wood shall be ~~of hot dipped zinc-coated galvanized steel, stainless steel, silicon bronze or copper~~ in accordance with this section. The coating weights for zinc-coated fasteners shall be in accordance with ASTM A 153.

R317.3.1 Fasteners for Preservative-treated Wood. Fasteners for preservative-treated wood shall be of hot dipped zinc-coated galvanized steel, stainless steel, silicon bronze or copper. Coating types and weights for connectors in contact with preservative-treated wood shall be in accordance with the connector manufacturer's recommendations. In the absence of manufacturer's recommendations, a minimum of ASTM A 653 type G185 zinc-coated galvanized steel, or equivalent, shall be used.

Exceptions:

1. One-half-inch (12.7 mm) diameter or ~~larger~~ greater steel bolts.

2. Fasteners other than nails and timber rivets shall be permitted to be of mechanically deposited zinc-coated steel with coating weights in accordance with ASTM B 695, Class 55 minimum.

R317.3 continues

R317.3 continued

<u>**R317.3.2 Fastenings for Wood Foundations.**</u>   <u>Fastenings for wood foundations shall be as required in AF&PA Technical Report No. 7.</u>

<u>**R317.3.3 Fasteners for Fire-retardant-treated Wood used in Exterior Applications or Wet or Damp Locations.**</u>   <u>Fasteners for fire-retardant-treated wood used in exterior applications or wet or damp locations shall be of hot dipped zinc-coated galvanized steel, stainless steel, silicon bronze or copper. Fasteners other than nails and timber rivets shall be permitted to be of mechanically deposited zinc-coated steel with coating weights in accordance with ASTM B 695, Class 55 minimum.</u>

<u>**R317.3.4 Fasteners for Fire-retardant-treated Wood used in Interior Applications.**</u>   <u>Fasteners for fire-retardant-treated wood used in interior locations shall be in accordance with the manufacturer's recommendations. In the absence of the manufacturer's recommendations, Section R317.3.3 shall apply.</u>

CHANGE SIGNIFICANCE: Changes to this section related to fasteners and connectors in contact with preservative-treated wood intend to clarify the applicable referenced standards and the minimum zinc coating weights for galvanized products. The standards are different for fasteners and connectors. Connectors include joist hangers, metal straps, and other metal products for connecting wood structural members, while fasteners typically are nails, screws and bolts. ASTM A 153, *Specification for Zinc Coating (Hot Dip) on Iron and Steel Hardware*, is the recognized standard for fasteners, which typically receive a coating of 1.0 ounce per square foot. ASTM A 653, *Specification for Steel Sheet, Zinc-coated (Galvanized) or Zinc-iron Alloy-coated (Galvanized) by the Hot-dip Process*, is the recognized standard for connectors manufactured from sheet goods. The current industry standard for connectors in contact with preservative-treated wood in a wet, damp or exterior location is ASTM A 653 type G60 or 0.60 ounce of zinc per square foot. The code now includes corrosion-resistance requirements for connectors in contact with preservative-treated wood in both interior and exterior locations that reflect the connector manufacturer's recommendations. If recommendations are not available, ASTM A 653 type G185, applies. The G185 designation indicates a coating thickness of 1.85 ounces of zinc per square foot or 3 times the current industry practice.

The other new provisions in Section R317 address fasteners for use with fire-retardant-treated wood (FRTW). The code now recognizes the manufacturer's recommendations for fasteners in contact with FRTW in interior applications. Fasteners in interior applications have not experienced corrosion problems compared with those in wet, damp, or exterior locations. Though moved to a new section for FRTW, the code still recognizes that fasteners appropriate for preservative-treated wood are also appropriate for FRTW in exterior, damp, or wet locations.

R317.4
Wood/Plastic Composites

CHANGE TYPE: Addition

CHANGE SUMMARY: A definition and specific requirements for manufactured wood/plastic composites are introduced into the IRC. Manufactured wood/plastic composite materials used in the construction of decks and other exterior structures must be labeled and listed as complying with ASTM D 7032 and installed according to the manufacturer's instructions. In addition to this new provision for protection against decay, the code places a reference to Section R317.4 in each of the applicable sections for stairs, handrails, guards, and deck boards where wood/plastic composites are used.

2009 CODE: **R317.4 Wood/Plastic Composites.** Wood/plastic composites used in exterior deck boards, stair treads, handrails and guardrail systems shall bear a label indicating the required performance levels and demonstrating compliance with the provisions of ASTM D 7032.

R317.4.1 Wood/plastic composites shall be installed in accordance with the manufacturer's instructions.

Section R202 Definitions
Wood/Plastic Composite. A composite material made primarily from wood or cellulose-based materials and plastic.

R311.7.4.4 Exterior Wood/plastic Composite Stair Treads. Wood/plastic composite stair treads shall comply with the provisions of Section R317.4.

R311.7.7.4 Exterior Wood/plastic Composite Handrails. Wood/plastic composite handrails shall comply with the provisions of Section R317.4.

R317.4 continues

R317.4 continued

R312.4 Exterior Wood/Plastic Composite Guards. Wood/plastic composite guards shall comply with the provisions of Section R317.4.

R502.1.7 Exterior Wood/Plastic Composite Deck Boards. Wood/plastic composites used in exterior deck boards shall comply with the provisions of Section R317.4.

R502.2.2.4 Exterior Wood/Plastic Composite Deck Boards. Wood/plastic composite deck boards shall be installed in accordance with the manufacturer's instructions.

CHANGE SIGNIFICANCE: Wood/plastic composite lumber is typically manufactured with a 50-50 mix of recovered wood sawdust and recycled plastics, though other combinations of wood or cellulose-based materials with various plastic materials may be used. Previously, wood/plastic composites were used with approval of the building official under the alternative materials and methods of construction provisions based on available data from the manufacturer and other sources such as ICC Evaluation Service (ES) reports. The 2009 IRC introduces a definition and prescriptive requirements for wood/plastic composites. The most common applications for wood/plastic composites are exterior decks and the associated stairs and railings. It follows that the requirements are appropriately placed in Section R317 for protection against decay. The referenced components in exterior locations must be labeled as complying with ASTM D 7032, *Standard Specification for Establishing Performance Ratings for Wood-Plastic Composite Deck Boards and Guardrail Systems (Guards or Handrails)*. The standard establishes spans for deck boards and deck boards used as stair treads to conform to the design load criteria of the code. It also recognizes guard and handrail characteristics to meet minimum code requirements. The installation of wood/plastic composite lumber, including the type of fasteners, must be according to the manufacturer's instructions.

CHANGE TYPE: Modification

CHANGE SUMMARY: When used for protection against termite damage, pressure-preservative-treated wood must now meet the location requirements for protection against decay in Section R317 in addition to the AWPA standards. A similar change applies to physical barriers, which previously did not identify any location requirements. Steel is now specifically recognized as being termite resistant. The definition of naturally resistant wood has been removed from Section R318 and revised definitions for naturally durable wood and termite-resistant material have been placed in Section R202.

2009 CODE: **R320.1** **R318.1 Subterranean Termite Control Methods.**   In areas subject to damage from termites as indicated by Table R301.2(1), methods of protection shall be one of the following methods or a combination of these methods:

1. Chemical termiticide treatment, as provided in Section R318.2.

2. Termite baiting system installed and maintained according to the label.

3. Pressure-preservative-treated wood in accordance with the ~~AWPA standards listed in Section R319.1~~ provisions of Section R317.1.

4. Naturally <u>durable</u> termite-resistant wood ~~as provided in Section R320.3~~.

5. Physical barriers as provided in Section R318.3 <u>and used in locations as specified in Section R317.1</u>

6. <u>Cold-formed steel framing in accordance with Sections R505.2.1 and R603.2.1.</u>

R318.1 continues

R318.1
Subterranean Termite Control Methods

R318.1 continued

R320.3 Naturally Resistant Wood. ~~Heartwood of redwood and eastern red cedar shall be considered termite resistant.~~

Section R202, Definitions

Naturally Durable Wood. The heartwood of the following species~~:~~ ~~Decay-resistant redwood, cedars, black locust and black walnut. Note:~~ with the exception that an occasional piece with corner sapwood is permitted if 90 percent or more of the width of each side on which it occurs is heartwood.

> **Decay resistant.** Redwood, cedar, black locust, and black walnut.
> **Termite resistant.** Alaska yellow cedar, redwood, eastern red cedar, and western red cedar, including all sapwood of western red cedar.

Termite-Resistant Material. Pressure-preservative-treated wood in accordance with the AWPA standards in Section R317.1, naturally durable termite-resistant wood, steel, concrete, masonry or other approved material.

CHANGE SIGNIFICANCE: References to Section R317 clarify the location requirements for termite protection in Section R318.1, Item 3, pressure-preservative-treated wood, and Item 5, physical barriers. Damage to wood from decay and from termites generally relates to its close proximity to the ground. Previously, the code provided no guidance as to the location of these termite protection methods.

The previous definition of naturally termite resistant wood recognized only the heartwood of redwood and eastern red cedar. This definition has been deleted from the text of Section R318, and the existing definition for naturally durable wood in Section R202 has been revised to distinguish termite resistant species from decay-resistant species to reflect current knowledge on the properties of the referenced species. The revised definition expands the wood species considered termite resistant to include Alaska yellow cedar and western red cedar, including all sapwood of western red cedar, in addition to redwood and eastern red cedar. A new definition has been added for termite-resistant material intending to clarify that in addition to pressure-preservative-treated wood and naturally durable wood, other approved materials that resist termites include steel, concrete, and masonry.

CHANGE TYPE: Modification

CHANGE SUMMARY: The IRC now prescribes the minimum size of address numbers and requires a contrasting background for visibility. To meet the legibility and visibility criteria for dwellings located a distance from the road, address numbers must be posted on a sign or monument near the road.

2009 CODE: ~~R321.1~~ R319.1 ~~Premises Identification.~~ Address Numbers. ~~Approved numbers or addresses shall be provided for all new buildings~~ Buildings shall have approved address numbers,

R319.1 continues

R319.1
Address Numbers

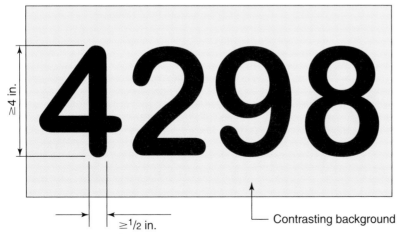

Minimum dimensions for address numbers

R319.1 continued   building numbers or approved building identification placed in ~~such~~ a position ~~as to be~~ that is plainly legible and visible from the street or road fronting the property. <u>These numbers shall contrast with their background. Address numbers shall be Arabic numbers or alphabetical letters. Numbers shall be a minimum of 4 inches (102 mm) high with a minimum stroke width of ½ inch (12.7 mm). Where access is by means of a private road and the building address cannot be viewed from the public way, a monument, pole or other sign or means shall be used to identify the structure.</u>

CHANGE SIGNIFICANCE: Previously, the code required address numbers for all new buildings to meet the performance criteria of being visible and legible from the road. The change prescribes numbers not less than 4 inches high with ½-inch-wide stroke and contrasting background. The new text intends to provide consistency with the *International Building Code* (IBC) and the *International Fire Code* (IFC) in adequately identifying buildings for emergency responders. In addition, this revision intends to clarify that identification, whether an address, building number, or other approved identification, applies to all buildings, not just new dwellings. The requirement for visibility and legibility from the public way remains. Additional text clarifies that for buildings located too far from the public way to meet these requirements, numbers must be mounted on a sign or monument sufficiently close to the street or road to pass the legibility and visibility test.

CHANGE TYPE: Modification

CHANGE SUMMARY: The code now directly references ASCE 24 for the design and construction of buildings or structures in floodways and coastal high-hazard V Zones. The prohibition against placement of fill under buildings in coastal V Zones has been deleted.

2009 CODE: R324.1 R322.1 General.   Buildings and structures constructed in whole or in part in flood hazard areas (including A or V Zones) as established in Table R301.2(1) shall be designed and constructed in accordance with the provisions contained in this section.

> **Exception:** Buildings and structures located in whole or in part in identified floodways shall be designed and constructed ~~as stipulated in the International Building Code~~ in accordance with ASCE 24.

R322.1.1 Alternative Provisions.   As an alternative to the requirements in Section R322.3 for buildings and structures located in whole or in part in coastal high hazard areas (V Zones), ASCE 24 is permitted subject to the limitations of this code and the limitations therein.

R322.1.4 Establishing the Design Flood Elevation.   The design flood elevation shall be used to define areas prone to flooding~~, and shall describe,~~. At a minimum, the design flood elevation is the higher of:

1. The base flood elevation at the depth of peak elevation of flooding (including wave height) which has a 1 percent (100-year flood) or greater chance of being equaled or exceeded in any given year, or
2. The elevation of the design flood associated with the area designated on a flood hazard map adopted by the community, or otherwise legally designated.

R322 continues

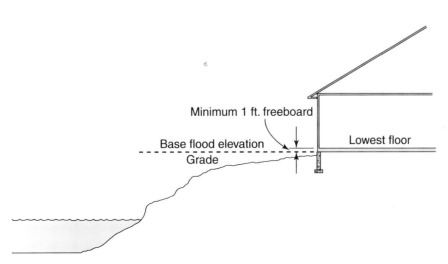

R322
R322
Flood-Resistant Construction

Flood resistant construction

R322 continued

R322.2 Flood Hazard Areas (Including A Zones). All areas that have been determined to be prone to flooding but not subject to high velocity wave action shall be designated as flood hazard areas. <u>Flood hazard areas that have been delineated as subject to wave heights between 1.5 feet (457 mm) and 3 feet (914 mm) shall be designated as Coastal A Zones.</u> All buildings and structures constructed in whole or in part in flood hazard areas shall be designed and constructed in accordance with Sections R322.2.1 through R322.2.3.

R322.3 Coastal High-hazard Areas (Including V Zones). (No change to text.)

R322.3.1 Location and site preparation.

1. ~~Buildings and structures~~ <u>New buildings and buildings that are determined to be substantially improved pursuant to Section R105.3.1.1</u> shall be located landward of the reach of mean high tide.

2. (No change to text.)

R322.3.2 Elevation Requirements.

1. All buildings and structures erected within coastal high hazard areas shall be elevated so that the lowest portion of all structural members supporting the lowest floor, with the exception of mat or raft foundations, piling, pile caps, columns, grade beams and bracing, is ~~located at or above the design flood elevation.~~:

   1.1. <u>Located at or above the design flood elevation, if the lowest horizontal structural member is oriented parallel to the direction of wave approach, where parallel shall mean less than or equal to 20 degrees from the direction of approach, or</u>

   1.2. <u>Located at the base flood elevation plus one foot (305 mm), or the design flood elevation, whichever is higher, if the lowest horizontal structural member is oriented perpendicular to the direction of wave approach, where perpendicular shall mean greater than 20 degrees from the direction of approach.</u>

2. Basement floors that are below grade on all sides are prohibited.

3. The use of fill for structural support is prohibited.

4. ~~The placement of fill beneath buildings and structures is prohibited.~~

4. <u>Minor grading, and the placement of minor quantities of fill, shall be permitted for landscaping and for drainage purposes under and around buildings, and for support of parking slabs, pool decks, patios and walkways.</u>

Because these code changes revised substantial portions of Section R322 (previously Section R324), the entire code change text is too extensive to be included here. Refer to Code Changes RB48-06/07, RB129-06/07, RB130-06/07, RB92–07/08, RB93–07/08, RB96–07/08, RB97–07/08, and RB100–07/08 in the *2009 IRC Code Changes Resource Collection* for the complete text and history of the code changes.

CHANGE SIGNIFICANCE: Certain parts of flood hazard areas have conditions (generally depth, velocity, wave heights, and potential debris impacts) that warrant design and construction according to the standard ASCE 24 *Flood Resistant Design and Construction.* Previously, the IRC referenced the IBC for design in these areas, and the IBC in turn, for the most part, referenced ASCE 24. The IRC now references ASCE 24 directly for design of buildings and structures in floodways, which are generally where flood velocities are higher and floodwaters are deeper. The code also provides an alternative to design to the same standard for buildings and structures in coastal high-hazard areas (V Zones), areas where wave heights of 3 feet or more are anticipated during the base flood. In many coastal communities, these areas also are subject to erosion and local scour.

The clarification regarding design flood elevation in Section R322.1.4 intends to provide consistency with the IBC and the referenced standard, ASCE 24 *Flood Resistant Design and Construction.* While the majority of communities use the flood maps prepared by the Federal Emergency Management Agency (FEMA), some communities and states prepare and adopt maps based on a different design flood—for example, using flood discharges based on anticipated development.

In coastal high-hazard areas (V Zone), the freeboard now specified in Section R322.3.2 matches ASCE 24 and is a function of whether the lowest horizontal structural member is parallel or perpendicular to the anticipated direction of wave approach.

While the placement of nonstructural fill can divert waves and increase the potential for scour around foundation elements, the intent of limiting fill is not to preclude placement of landscaping materials or replacement of sand and soil that may be removed during flood conditions. Therefore, the prohibition against fill beneath buildings in high-hazard coastal areas was deemed overly restrictive and has been removed.

R323
Storm Shelters

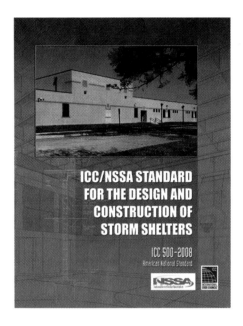

CHANGE TYPE: Addition

CHANGE SUMMARY: This new section provides that storm shelters, when built or installed, must be constructed in accordance with the new ICC/NSSA-500 *Standard on the Design and Construction of Storm Shelters.*

2009 CODE: **Section R323 Storm Shelters**

R323.1 General.   This section applies to the construction of storm shelters when constructed as separate detached buildings or when constructed as safe rooms within buildings for the purpose of providing safe refuge from storms that produce high winds, such as tornados and hurricanes. In addition to other applicable requirements in this code, storm shelters shall be constructed in accordance with ICC/NSSA-500.

CHAPTER 44 REFERENCED STANDARDS
ICC
ICC 500-08 ICC/NSSA *Standard on the Design and Construction of Storm Shelters*

CHANGE SIGNIFICANCE: The 2009 IRC does not require storm shelters, but when installed, they must be constructed in accordance with the new ICC/NSSA-500 *Standard on the Design and Construction of Storm Shelters.* Storm shelters intend to protect occupants from serious injury from flying debris in the event of a tornado, hurricane, or other high-wind event while maintaining a minimum interior environment. Shelters conforming to the ICC-500 standard are designed to withstand impact from windborne projectiles, referred to as missiles, such as a 2 by 4 or other construction or natural material debris, that are common to high-wind events. The outer shell of above-ground shelters may be concrete, steel, composite materials, or other approved materials that have been tested to the prescribed large missile tests. Whether prefabricated or site built, storm shelters may be detached from the dwelling, either below or above ground, or installed within a dwelling in a basement or on the main floor. The International Code Council (ICC) and the National Storm Shelter Association (NSSA) jointly developed the referenced consensus standard, ICC/NSSA-500.

CHANGE TYPE: Modification

CHANGE SUMMARY: Where it is not feasible to provide the prescribed fall of 6 inches within the first 10 feet away from a foundation, the code includes new performance language requiring drainage away from the foundation without prescribing a slope.

2009 CODE: **R401.3 Drainage.** Surface drainage shall be diverted to a storm sewer conveyance or other approved point of collection that does not create a hazard. Lots shall be graded to drain surface water away from foundation walls. The grade shall fall a minimum of 6 inches (152 mm) within the first 10 feet (3048 mm).

> **Exception:** Where lot lines, walls, slopes or other physical barriers prohibit 6 inches (152 mm) of fall within 10 feet (3048 mm), ~~the final grade shall slope away from the foundation at a minimum slope of 5 percent and the water shall be directed to~~ drains or swales <u>shall be constructed</u> to ensure drainage away from the structure. ~~Swales shall be sloped a minimum of 2 percent when located within 10 feet (3048 mm) of the building foundation.~~ Impervious surfaces within 10 feet (3048 mm) of the building foundation shall be sloped a minimum of 2 percent away from the building.

CHANGE SIGNIFICANCE: The IRC prescribes methods to direct surface water away from the foundation to an approved location. Proper design of surface drainage prevents water intrusion into basements and crawl spaces, potential damage of building materials, increased

R401.3
Surface Drainage

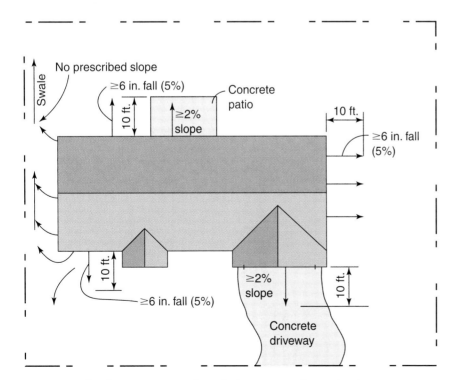

Grade to ensure surface drainage away from structure

R401.3 continues

R401.3 continued lateral pressure on the foundation, and nuisance ponding on the lot. For dwellings placed on small lots or lots with other impediments to surface drainage, it is not always feasible to provide the prescribed fall of 6 inches within the first 10 feet (5 percent average slope) away from the foundation. The 2006 IRC afforded an alternative to provide a 5 percent slope adjacent to the foundation and to further direct the surface drainage with swales sloped not less than 2 percent. Depending on the existing slope and configuration of the lot, such slopes may also be difficult to achieve. The revised exception provides performance language to achieve surface drainage without specifying minimum slopes. The intent is to allow as moderate a grade as possible to prevent slope instability and erosion and still drain surface water to an approved location. The performance criteria recognizes that the appropriate slope for the lot is a function of the combined ground frost and moisture conditions, soil type, geological conditions, and local geographic conditions. The 2009 IRC maintains the minimum slope provisions of 2 percent (¼ inch per foot) for impervious surfaces within 10 feet of the building. This matches the industry standard practice for exterior concrete flatwork.

CHANGE TYPE: Modification

CHANGE SUMMARY: The revised text defines the necessary criteria for requiring a soil test in more objective terms based on available scientific data.

2009 CODE: **R401.4 Soil Tests.** ~~In areas likely to have~~ Where quantifiable data created by accepted soil science methodologies indicate expansive, compressible, shifting or other ~~unknown~~ questionable soil characteristics are likely to be present, the building official shall determine whether to require a soil test to determine the soil's characteristics at a particular location. This test shall be made by an approved agency using an approved method.

CHANGE SIGNIFICANCE: The previous language giving authority to require a soil test was deemed subjective and open to various interpretations. At issue was the phrase "in areas likely to have," which gave no indication what criteria were to be used to determine whether the unsuitable soils were likely to occur at a given site, possibly allowing a jurisdiction to require a soil test for all building sites without justification. The revised text requires that a determination be based on existing soil maps, test data records, or other documentation with quantifiable data that are based on accepted geotechnical methodologies. When the data exist, the code directs the building official to make a determination of whether to require soil testing or not.

R401.4
Soil Tests

R402.3

Precast Concrete Foundation Materials

CHANGE TYPE: Modification

CHANGE SUMMARY: Minimum specifications for materials used in the manufacture of precast concrete foundations have been added to the code. The reference to Section R404.6 points the user to new provisions for the design and installation of precast concrete foundations.

2009 CODE: R402.3 Precast Concrete. ~~Approved~~ Precast concrete foundations shall be designed in accordance with Section R404.5 and shall be installed in accordance with the provisions of this code and the manufacturer's installation instructions.

R402.3.1 Precast Concrete Foundation Materials. Materials used to produce precast concrete foundations shall meet the following requirements.

1. All concrete used in the manufacture of precast concrete foundations shall have a minimum compressive strength of 5,000 psi (34 470 kPa) at 28 days. Concrete exposed to a freezing and thawing environment shall be air entrained with a minimum total air content of 5 percent.

2. Structural reinforcing steel shall meet the requirements of ASTM A615, A706, or A996. The minimum yield strength of reinforcing steel shall be 40,000 psi (Grade 40) (276 MPa). Steel reinforcement for precast concrete foundation walls shall have a minimum concrete cover of ¾ inch (19.1 mm).

3. Panel-to-panel connections shall be made with Grade II steel fasteners.

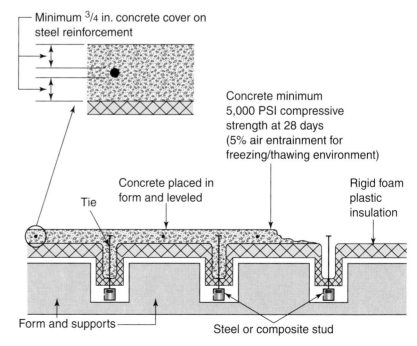

Example of precast foundation wall manufacturing

4. The use of nonstructural fibers shall conform to ASTM C 1116.

5. Grout used for bedding precast foundations placed upon concrete footings shall meet ASTM C 1107.

CHANGE SIGNIFICANCE: Precast concrete foundations were first specifically recognized in the IRC in the 2003 edition. Manufactured in a controlled environment and delivered to the jobsite, these engineered systems are set on crushed stone or concrete footings in accordance with the manufacturer's instructions. The 2009 IRC establishes minimum requirements for the materials used in the manufacture of precast concrete foundations. When compared with cast-in-place concrete, precast concrete requires a higher specified compressive strength of 5000 psi at 28 days curing time. The greater strength, achieved in factory-controlled conditions for forming, temperature, water content, mixing, placing, and curing, is necessary to accommodate the desired thickness and profile of the engineered components. Section R402.3 now references a new Section R404.5 for the design and installation provisions for precast concrete foundation walls.

R403.1.3.2

Seismic Reinforcing for Slabs-on-Ground with Turned-Down Footings

CHANGE TYPE: Modification

CHANGE SUMMARY: For turned-down footings in Seismic Design Categories (SDCs) D_0, D_1, and D_2, this change clarifies that the exception permitting bars in the middle of the footing depth is an alternative to the top and bottom bar location of horizontal reinforcing in the footing. When the concrete for the footing is placed prior to the placement of the slab, vertical dowels with hooks are required to connect the slab to the footing.

2009 CODE: R403.1.3.2 Slabs-on-Ground with Turned-Down Footings. Slabs-on-ground with turned-down footings shall have a minimum of one No. 4 bar at the top and the bottom of the footing.

> **Exception:** For slabs-on-ground cast monolithically with the footing, locating one No. 5 bar or two No. 4 bars in the middle third of the footing depth shall be permitted located in the middle third of the footing depth as an alternative to placement at the footing top and bottom.

Where the slab is not cast monolithically with the footing, No. 3 or larger vertical dowels with standard hooks on each end shall be provided in accordance with Figure R403.1.3.2. Standard hooks shall comply with Section R611.5.4.5.

CHANGE SIGNIFICANCE: Section R403.1.3.2 applies to the reinforcing of concrete slabs with turned-down footings in Seismic Design Categories (SDCs) D_0, D_1, and D_2. The footings generally require a minimum of one No. 4 horizontal reinforcing bar continuous at the top and one No. 4 horizontal reinforcing bar continuous at the bottom of the footing. The exception permits greater reinforcing—one No. 5 bar or two No. 4 bars—located in the middle third of the footing depth, but

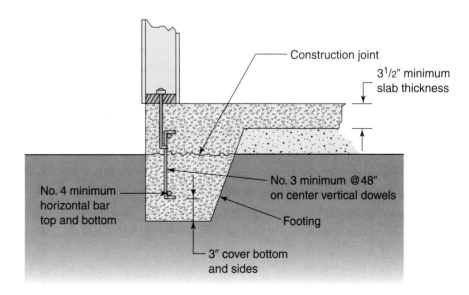

Dowels for slabs-on-ground with turned-down footings

only if the slab is cast monolithically with the footing. That is, the concrete is placed in one continuous piece without seams or joints between the footing and the slab. To address the possibility that a code user might infer that reinforcing must be located in the middle third where the footing and slab are monolithic, the new wording clarifies that the exception is an alternative. Though the IRC permits reinforcing in the center third, placement at the top and bottom is considered to provide better crack control and stiffness, and remains as an option.

When the concrete slab-on-ground and the turned-down footing are not cast monolithically, and the edge of the slab is placed over the footing, a horizontal construction joint occurs where the slab rests on the footing. Based on observation of damage due to slippage in earthquake events, the 2009 IRC requires vertical No. 3 reinforcing bars with standard hooks at each end to be installed at 48 inches on center to reinforce this horizontal construction joint.

R403.1.6
Foundation Anchorage

CHANGE TYPE: Modification

CHANGE SUMMARY: The revision and reorganization of Section R403.1.6 removes redundant language and clarifies the anchorage requirements for wood sill and sole plates resting on concrete and masonry foundations. Wood bottom plates of exterior walls, bottom plates of interior braced walls, and all wood sill plates require anchor bolts spaced a maximum of 6 feet on center. The code no longer allows wood plate anchorage to brick or solid masonry foundations. Anchor bolts must be placed in concrete or in the grouted cells of hollow concrete masonry units (CMUs). The bolting requirement for cold-formed steel bottom track has been removed in favor of references to applicable requirements for cold-formed steel framing.

2009 CODE: R403.1.6 Foundation Anchorage. ~~When braced~~ <u>Sill plates and</u> wall<u>s</u> ~~panels are~~ supported directly on continuous foun-

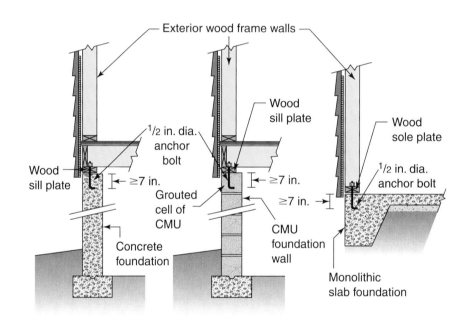

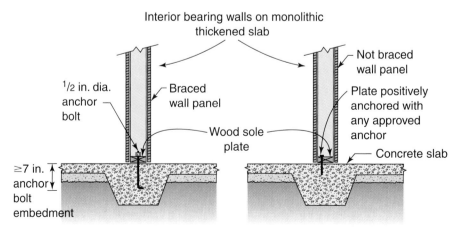

Foundation anchorage

dations, ~~the wall wood sill plate or cold-formed steel bottom track~~ shall be anchored to the foundation in accordance with this section.

~~The~~ Wood sole plate<u>s</u> at <u>all</u> exterior walls on monolithic slabs, <u>wood sole plates of braced wall panels at building interiors on monolithic slabs</u>, and <u>all</u> wood sill plates shall be anchored to the foundation with anchor bolts spaced a maximum of 6 feet (1829 mm) on center. <u>Bolts shall be at least ½ inch (12.7 mm) in diameter and shall extend a minimum of 7 inches (178 mm) into concrete or grouted cells of concrete masonry units. A nut and washer shall be tightened on each anchor bolt.</u> There shall be a minimum of two bolts per plate section with one bolt located not more than 12 inches (305 mm) or less than seven bolt diameters from each end of the plate section. ~~In Seismic Design Categories D₀, D₁ and D₂, anchor bolts shall be spaced at 6 feet (1829 mm) on center and located within 12 inches (305 mm) of the of each plate section at interior braced wall lines when required by Section R602.10.9 to be supported on a continuous foundation. Bolts shall be at least 1/2 inch (13 mm) in diameter and shall extend a minimum of 7 inches (178 mm) into masonry or concrete.~~ Interior bearing wall sole plates on monolithic slab foundation <u>that are not part of a braced wall panel</u> shall be positively anchored with approved fasteners. ~~A nut and washer shall be tightened on each bolt of the plate.~~ Sill<u>s</u> <u>plates</u> and sole plates shall be protected against decay and termites where required by Sections R317 and R318. Cold-formed steel framing systems shall be fastened to the wood sill plates or anchored directly to the foundation as required in Section R505.3.1 or R603.1.1.

Exceptions:
1. Foundation anchorage, spaced as required to provide equivalent anchorage to ½-inch-diameter (13 mm) anchor bolts.

2. Walls 24 inches (610 mm) total length or shorter connecting offset braced wall panels shall be anchored to the foundation with a minimum of one anchor bolt located in the center third of the plate section and shall be attached to adjacent braced wall panels <u>at corners as shown in</u> ~~per~~ Figure ~~R602.10.5~~ <u>R602.10.4.4(1)</u> ~~at corners~~.

3. Walls 12 inches (305 mm) total length or shorter connecting offset braced wall panels shall be permitted to be connected to the foundation without anchor bolts. The wall shall be attached to adjacent braced wall panels <u>at corners as shown in</u> ~~per~~ Figure ~~R602.10.5~~ <u>R602.10.4.4(1)</u> ~~at corners~~.

CHANGE SIGNIFICANCE: Section R403.1.6 has been rearranged to clarify the foundation anchorage requirements, particularly as they relate to wall bracing of conventional wood frame construction. In wood frame construction vernacular, a sill plate is a horizontal wood member that rests on the foundation. A sole plate is a horizontal wood member that serves as the base for the studs in a wall or partition. In raised floor construction, there is a sill plate on the foundation and a sole plate above on the floor sheathing that supports the studs. In a slab on grade foundation, the plate rests on the foundation and serves as the base for studs, so it can be called a sill plate or a sole plate. In

R403.1.6 continues

R403.1.6 continued

the language of the IRC, the plate on a slab on grade foundation is referred to as a sole plate. The intent of the code is that, where supported directly on continuous foundations, sill plates, sole plates at all exterior walls, and sole plates of braced wall panels at building interiors are required to be anchored to the foundation in accordance with the section. Typically this is achieved through the use of the prescribed anchor bolts spaced not more than 6 feet on center. For bottom plates of an interior bearing wall located on a foundation, the code permits anchorage with any approved fasteners, provided the portion of wall is not a braced wall panel. In this case, the code does not specify the type, size, or spacing of the fasteners, provided they are approved by the building official.

It is important to note that this section applies to anchorage of wood sill and sole plates to continuous foundations. Monolithic slab, in this context, means that the turned-down or thickened portions of the slab form a continuous footing that is integral with the concrete slab.

For sill plates anchored to masonry foundation walls, the code now specifically requires the anchor bolt to be embedded in grout filling the hollow core of a CMU, also referred to as concrete block. This change reflects a consensus that anchor bolts installed into the mortar joints of brick or other solid masonry products do not have sufficient load-carrying capacity to resist shear loads from braced wall panels.

The requirements for anchor bolt spacing in Seismic Design Categories D_0, D_1, and D_2 appear in Section R403.1.6.1. Duplication of those requirements in Section R403.1.6 was therefore deemed unnecessary and has been deleted. The deletion of cold-formed steel bottom track from the charging statement of this section does not lessen the anchorage requirements for cold-formed steel framing. The applicable sections of R505.3.1 and R603.1.1 are still referenced for the anchorage requirements for cold-formed steel framing systems. In addition, such systems must conform to the requirements of AISI S230, *Standard for Cold-Formed Steel Framing—Prescriptive Method for One- and Two-Family Dwellings.*

CHANGE TYPE: Addition

CHANGE SUMMARY: Prescriptive requirements for crushed stone footings supporting precast concrete foundations are now included in the code.

2009 CODE: R403.4 Footings for Precast Concrete Foundations.   Footings for precast concrete foundations shall comply with Section R403.4.

R403.4.1 Crushed Stone Footings.   Clean crushed stone shall be free from organic, clayey or silty soils. Crushed stone shall be angular in nature and meet ASTM C 33, with the maximum size stone not to exceed ½ inch (12.7 mm) and the minimum stone size not to be smaller than $1/16$-inch (1.6 mm). Crushed stone footings for precast foundations shall be installed in accordance with Figure R403.4 (1)

R403.4

Footings for Precast Concrete Foundations

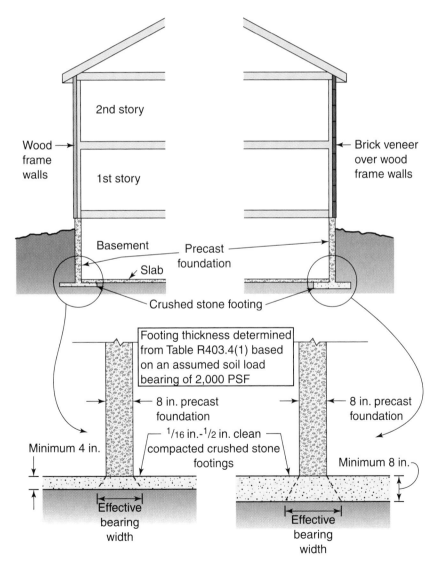

2nd story

Wood frame walls

1st story

Brick veneer over wood frame walls

Basement   Precast foundation

Slab

Crushed stone footing

Footing thickness determined from Table R403.4(1) based on an assumed soil load bearing of 2,000 PSF

8 in. precast foundation

8 in. precast foundation

Minimum 4 in.

$1/16$ in.-$1/2$ in. clean compacted crushed stone footings

Minimum 8 in.

Effective bearing width

Effective bearing width

Minimum depth of crushed stone footings supporting precast foundations in seismic design categories A, B, and C

R403.4 continues

R403.4 continued

and Table R403.4. Crushed stone footings shall be consolidated using a vibratory plate in a maximum of 8-inch lifts. Crushed stone footings shall be limited to Seismic Design Categories A, B and C.

R403.4.2 Concrete Footings. Concrete footings shall be installed in accordance with Section R403.1 and Figure R403.4 (2).

CHANGE SIGNIFICANCE: Precast concrete foundations were first specifically recognized in the 2003 IRC. Installation of these engineered systems must follow the manufacturer's instructions, including provisions for adequate footings. Previously, the code prescribed solid or fully grouted masonry or concrete footings, or other approved footing systems. Crushed stone footings are commonly used for both wood foundations and precast foundations and have performed well for many years. Prior to the 2009 edition, the IRC permitted the installation of crushed stone footings for precast foundations when approved by the building official, who would base the approval on the manufacturer's information and other data. With this change, the code prescribes the materials, dimensions, and methods of installation for crushed stone footings supporting precast concrete foundations. The material specifications are similar to those for wood foundation footings.

For precast foundation footings, the code requires placement of the crushed stone in maximum 8-inch lifts with vibratory plate compaction. Minimum footing width is not specified other than to follow the manufacturer's instructions. In practice, crushed stone footings often project 24 inches or more beyond the faces of the foundation wall on both sides and may also serve as part of the sub-slab or foundation drainage system. In transferring the building loads to the soil, the ef-

TABLE R403.4 Minimum Depth of Crushed Stone Footings, D (inches)

		Load Bearing Value of Soil (psf)														
		1500				2000				3000				4000			
		MH, CH, CL, ML				SC, GC, SM, GM, SP, SW				GP, GW							
		Wall Width (inches)				Wall Width (inches)				Wall Width (inches)				Wall Width (inches)			
		6	8	10	12	6	8	10	12	6	8	10	12	6	8	10	12
Conventional Light-Frame Construction																	
1-Story	(1100 plf)	6	4	4	4	6	4	4	4	6	4	4	4	6	4	4	4
2-Story	(1800 plf)	8	6	4	4	6	4	4	4	6	4	4	4	6	4	4	4
3-Story	(2000 plf)	16	14	12	10	10	8	6	6	6	4	4	4	6	4	4	4
4-Inch Brick Veneer over Light-Frame or 8-Inch Hollow Concrete Masonry																	
1-Story	(1500 plf)	6	4	4	4	6	4	4	4	6	4	4	4	6	4	4	4
2-Story	(2700 plf)	14	12	10	8	10	8	6	4	6	4	4	4	6	4	4	4
3-Story	(4000 plf)	22	22	20	18	16	14	12	10	10	8	6	4	6	4	4	4
8-inch Solid or Fully Grouted Masonry																	
1-Story	(2000 plf)	10	8	6	4	6	4	4	4	6	4	4	4	6	4	4	4
2-Story	(3600 plf)	20	18	16	16	14	12	10	8	8	6	4	4	6	4	4	4
3-Story	(5300 plf)	32	30	28	26	22	22	20	18	14	12	10	8	10	8	6	4

fective bearing width of the footing is determined by the width of the foundation wall and the depth of the stone footing. As the depth of the crushed stone increases, the effective bearing width on the underlying soil also increases. Therefore, the IRC prescribes the minimum footing depth based on the foundation wall width, the load-bearing value of the soil, the height of the building being supported, and the type of construction.

Crushed stone footings for precast concrete foundations are not allowed for buildings sited in Seismic Design Categories D_0, D_1, and D_2. Requirements for concrete footings supporting precast concrete foundation walls match those for masonry and cast-in-place concrete foundation walls. Installation of continuous non-shrink grout is required between the top of the concrete footing and the base of the precast concrete foundation.

Tables R404.1(1) through R404.1(3)

Lateral Support for Concrete and Masonry Foundation Walls

CHANGE TYPE: Deletion

CHANGE SUMMARY: The prescriptive lateral restraint provisions for the top of concrete and masonry foundation walls based on soil type, height of wall, and unbalanced backfill height have been removed from the code.

2009 CODE: **R404.1 Concrete and Masonry Foundation Walls.**

Delete second paragraph without substitution:
~~Foundation walls that meet all of the following shall be considered laterally supported:~~

1. ~~Full basement floor shall be 3.5 inches (89 mm) thick concrete slab poured tight against the bottom of the foundation wall.~~

2. ~~Floor joists and blocking shall be connected to the sill plate at the top of wall by the prescriptive method called out in Table R404.1(1), or; shall be connected with an approved connector with listed capacity meeting Table R404.1(1).~~

3. ~~Bolt spacing for the sill plate shall be no greater than per Table R404.1(2).~~

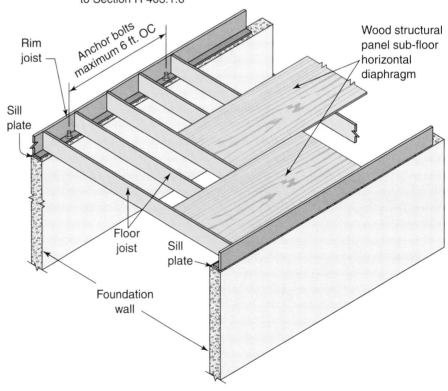

Connections and fasteners for wood framing must conform to Table R602.3(1) Foundation anchorage must conform to Section R 403.1.6

Rim joist

Anchor bolts maximum 6 ft. OC

Wood structural panel sub-floor horizontal diaphragm

Sill plate

Floor joist

Sill plate

Foundation wall

Prescriptive lateral support for top of foundation walls

4. ~~Floor shall be blocked perpendicular to the floor joists. Blocking shall be full depth within two joist spaces of the foundation wall, and be flat-blocked with minimum 2-inch by 4-inch (51 mm by 102 mm)blocking elsewhere.~~

5. ~~Where foundation walls support unbalanced load on opposite sides of the building, such as a daylight basement, the building aspect ratio, L/W, shall not exceed the value specified in Table R404.1(3). For such foundation walls, the rim board shall be attached to the sill with a 20 gage metal angle clip at 24 inches (610 mm) on center, with five 8d nails per leg, or an approved connector supplying 230 pounds per linear foot (3.36 kN/m) capacity.~~

~~**Table R404.1(1)**~~
~~**Top Reactions and Prescriptive Support For Foundation Walls**[a]~~

~~**Table R404.1(2)**~~
~~**Maximum Plate Anchor-Bolt Spacing For Supported Foundation Wall**[a]~~

~~**Table R404.1(3)**~~
~~**Maximum Aspect Ratio, L/W for Unbalanced Foundations**~~

CHANGE SIGNIFICANCE: Prior to the 2006 edition, the IRC and the legacy *CABO One- and Two-Family Dwelling Code* relied on the prescriptive provisions for anchor bolt spacing and floor framing connections to provide lateral restraint at the top of basement foundation walls. As the result of engineering analysis, the 2006 IRC introduced additional requirements for anchor bolt spacing, floor joist attachment and blocking, and maximum aspect ratio for walk-out basements. The tabular values for these new requirements varied based on the height of the foundation wall, the maximum unbalanced backfill height, and the type of soil. Deletion of these lateral restraint provisions brings the 2009 IRC into agreement with the 2000 and 2003 editions.

Proponents of removing the top of foundation wall lateral restraint provisions reasoned that the traditional prescriptive provisions for anchor bolts and floor system connections have performed well for many years without substantiated problems or failures. It was also argued that the engineering analysis used to justify introduction of the lateral restraint provisions was overly conservative and failed to take into account all design factors.

R404.1

Concrete and Masonry Foundation Walls

CHANGE TYPE: Modification

CHANGE SUMMARY: The technical provisions for concrete foundation walls have been separated from the masonry foundation provisions. Insulated concrete form (ICF) foundation wall requirements have been integrated with the conventional concrete foundation wall provisions. The 2009 IRC revises the prescriptive concrete foundation wall requirements to reflect the provisions of the new referenced Portland Cement Association standard PCA 100 *Prescriptive Design of Exterior Concrete Walls for One- and Two-Family Dwellings.*

2009 CODE: **R404.1 Concrete and Masonry Foundation Walls.** <u>Concrete foundation walls shall be selected and constructed in accordance with the provisions of Section R404.1.2.</u> and m Masonry foundation walls shall be selected and constructed in accordance with the provisions of Section R404<u>.1.1</u> ~~or in accordance with ACI 318, ACI 332, NCMATR68–A or ACI530/ASCE 5/TMS 402 or other approved structural standards. When ACI 318, ACI 332 or ACI 530/ASCE 5/TMS 402 or the provisions of Section R404 are used to design concrete or masonry foundation walls, project drawings, typical details and specifications are not required to bear the seal of the architect or engineer responsible for design, unless otherwise required by the state law of the jurisdiction having authority.~~

R404.1.1 Design of Masonry Foundation Walls. <u>Masonry foundation walls shall be designed and constructed in accordance with the provisions of this section or in accordance with the provisions of ACI530/ASCE 5/TMS 402 or NCMA TR68-A . When ACI530/ASCE 5/TMS 402, NCMA TR68-A, or the provisions of this section are used to design masonry foundation walls, project drawings, typical details, and specifications are not required to bear the seal of the architect or engineer responsible for design, unless otherwise required by the state law of the jurisdiction having authority.</u>

R404.1.1<u>.1</u> Masonry Foundation Walls. Concrete masonry and clay masonry foundation walls shall be constructed as set forth in Table R404.1.1(1), R404.1.1(2), R404.1.1(3) or R404.1.1(4) and shall also comply with ~~the provisions of Section R404 and~~ the applicable provisions of Sections R606, R607 and R608. In <u>buildings assigned to</u> Seismic Design Category D_0, D_1 or D_2, concrete masonry and clay masonry foundation walls shall also comply with Section R404.1.4<u>.1</u>. Rubble stone masonry foundation walls shall be constructed in accordance with Sections R404.1.8 and R607.2.2. Rubble stone masonry walls shall not be used in buildings assigned to Seismic Design Category D_0, D_1 or D_2.

R404.1.2 Concrete Foundation Walls. ~~Concrete foundation walls shall be constructed as set forth in Table R404.1.1(5) and shall also comply with the provisions of Section R404 and the applicable provisions of Section R402.2. In Seismic Design Categories D_0, D_1 or D_2, concrete foundation walls shall also comply with Section R404.1.4.~~

Concrete foundation walls that support light-frame walls shall be designed and constructed in accordance with the provisions of this section, ACI 318, ACI 332 or PCA 100. Concrete foundation walls that support above-grade concrete walls that are within the applicability limits of Section R611.2 shall be designed and constructed in accordance with the provisions of this section, ACI 318, ACI 332, or PCA 100. Concrete foundation walls that support above-grade concrete walls that are not within the applicability limits of Section R611.2 shall be designed and constructed in accordance with the provisions of ACI 318, ACI 332, or PCA 100. When ACI 318, ACI 332, PCA 100 or the provisions of this section are used to design concrete foundation walls, project drawings, typical details and specifications are not required to bear the seal of the architect or engineer responsible for design, unless otherwise required by the state law of the jurisdiction having authority.

R404.1.2.1 Concrete Cross-section. Concrete walls constructed in accordance with this code shall comply with the shapes and minimum concrete cross-sectional dimensions required by Table R611.3. Other types of forming systems resulting in concrete walls not in compliance with this section and Table R611.3 shall be designed in accordance with ACI 318.

R404.4 Insulating Concrete Form Foundation Walls. ~~Insulating concrete form (ICF) foundation walls shall be designed and constructed in accordance with the provisions of this section or in accordance with the provisions of ACI 318. When ACI 318 or the provisions of this section are used to design insulating concrete form foundation walls, project drawings, typical details and specifications are not required to bear the seal of the architect or engineer responsible for design unless otherwise required by the state law of the jurisdiction having authority.~~

Because this code change deleted, added, and modified substantial portions of Section R404.1, the entire code change text is too extensive to be included here. Refer to Code Change RB116-07/08 in the *2009 IRC Code Changes Resource Collection* for the complete text and history of the code change.

CHANGE SIGNIFICANCE: The concrete foundation wall provisions have been substantially revised and are now separated from the masonry foundation wall provisions. There are no technical changes to the masonry provisions. This change also correlates and integrates the insulating concrete form (ICF) provisions into the conventionally formed cast-in-place concrete provisions. This arrangement is preferred to maintaining the provisions in separate sections and intends to provide consistency in the application of the code. The Portland Cement Association has developed a new consensus standard PCA 100 *Prescriptive Design of Exterior Concrete Walls for One- and Two-Family Dwellings* that is now referenced by the IRC. The prescriptive concrete provisions of Section R404.1 are based on PCA 100, and the tabular values for vertical reinforcement are revised to reflect changes to the referenced standards ACI 318 and ASCE 7. In addition to the

R404.1 continues

R404.1 continued

provisions of the referenced standards ACI 318 and ACI 332, PCA 100 is referenced as another option for alternate design of concrete foundation walls that are beyond the scope of the prescriptive provisions of the IRC.

With these changes, the prescriptive provisions for concrete foundation walls are more comprehensive. New provisions govern the location, cover, and continuity of reinforcement, lap splices and standard hooks, and installation of construction joints. The provisions also incorporate technical requirements for constructing concrete stem wall foundations not presently in the code. In the 2006 IRC, the tables for removable form concrete walls required a yield strength of 60 000 psi (Grade 60) reinforcing steel, and the tables for ICF walls required 40 000 psi (Grade 40) steel. The vertical reinforcement tables for both types of walls are now based on reinforcement steel with a yield strength of 60 000 psi (Grade 60). However, a new table provides more flexibility for the use of different bar sizes or grades of steel than specified in the other tables. In addition, the 2009 IRC now specifies the material and placement requirements for concrete mixing, delivery, aggregate size, proportioning, slump, and consolidation (vibration). New requirements also specify approved materials for forms and form ties.

CHANGE TYPE: Addition

CHANGE SUMMARY: This new section in the IRC requires engineering and sets design and labeling requirements for precast foundation walls. Design drawings must be prepared by a registered design professional, include the applicable design criteria and be submitted to the building official for approval.

2009 CODE: R404.5 Precast Concrete Foundation Walls.
R404.5.1 Design. Precast concrete foundation walls shall be designed in accordance with accepted engineering practice. The design and manufacture of precast concrete foundation wall panels shall comply with the materials requirements of Section R402.3 or ACI 318. The panel design drawings shall be prepared by a registered design professional where required by the statutes of the jurisdiction in which the project is to be constructed in accordance with Section R106.1.

R404.5.2 Precast Concrete Foundation Design Drawings.
Precast concrete foundation wall design drawings shall be submitted to the building official and approved prior to installation. Drawings shall include, at a minimum, the information specified below:

1. Design loading as applicable
2. Footing design and material
3. Concentrated loads and their points of application
4. Soil-bearing capacity
5. Maximum allowable total uniform load
6. Seismic Design Category
7. Basic wind speed

R404.5
Precast Concrete Foundation Walls

R404.5 continues

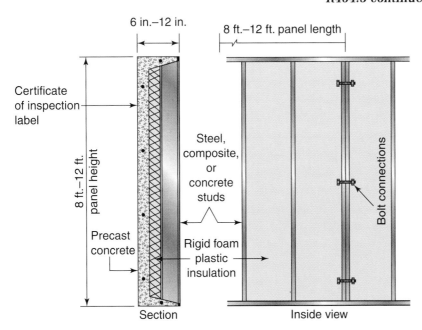

Example of precast concrete foundation wall

R404.5 continued

R404.5.3 Identification. Precast concrete foundation wall panels shall be identified by a certificate of inspection label issued by an approved third party inspection agency.

CHANGE SIGNIFICANCE: The addition of Section R404.5 intends to clarify requirements for the design of precast concrete foundation wall systems. The new provisions require a design in accordance with accepted engineering practice, recognized engineering standards and submittal of design drawings to the building official for approval. Third-party inspection and placement of certificate-of-inspection labels on the precast panels verify that the product is manufactured in a plant to code-prescribed requirements under verified quality control.

Precast foundation systems are engineered products based on several design approaches including, but not limited to, stud and cavity, solid wall panel, composite panel, and hollow core systems. The minimum performance design criteria in Section R404.5 do not favor or exclude any specific system, providing neutral and nonproprietary requirements.

CHANGE TYPE: Addition

CHANGE SUMMARY: This new section is specific to foundation drainage requirements for precast foundations supported by crushed stone footings. Drainage pipe must be installed a minimum of 1 foot beyond the edge of the wall to preserve the integrity of the effective bearing surface of the crushed stone footing.

2009 CODE: <u>**R405.1.1 Precast Concrete Foundation.** Precast concrete walls that retain earth and enclose habitable or useable space located below grade that rest on crushed stone footings shall have a perforated drainage pipe installed below the base of the wall on either the interior or exterior side of the wall, at least one foot (305 mm) beyond the edge of the wall. If the exterior drainage pipe is used, an approved filter membrane material shall cover the pipe. The drainage system shall discharge into an approved sewer system or to daylight.</u>

CHANGE SIGNIFICANCE: Concrete and masonry foundations enclosing spaces below grade, typically basements, require a means to drain water away from the base of the foundation. These drainage requirements also apply to precast concrete foundations. The addition of this subsection further stipulates that precast concrete foundations with crushed stone footings specifically require a perforated drainage pipe located below the base of the wall and not less than one foot from the exterior or interior face of the wall. The type and location of the drainage pipe is instrumental in preserving the bearing capabilities of the crushed stone footing.

R405.1.1
Precast Concrete Foundation Drainage

Precast concrete foundation drainage

R406.4

Precast Concrete Foundation System Dampproofing

CHANGE TYPE: Addition

CHANGE SUMMARY: Precast concrete basement foundations require panel joints to be filled and sealed and the exterior below-grade surface to be dampproofed to prevent water intrusion into the below-grade space.

2009 CODE: **R406.4 Precast Concrete Foundation System Dampproofing.** Except where required by Section R406.2 to be waterproofed, precast concrete foundation walls enclosing habitable or useable spaces located below grade shall be dampproofed in accordance with Section R406.1.

R406.4.1 Panel Joints Sealed. Precast concrete foundation panel joints shall be sealed full height with a sealant meeting ASTM C 920, Type S or M, Grade NS, Class 25, Use NT, M or A. Joint sealant shall be installed in accordance with the manufacturer's installation instructions.

CHANGE SIGNIFICANCE: Precast concrete foundations were first specifically recognized in the 2003 edition of the IRC. Such foundations are an engineered system of panels that are set on the footing and connected at the panel joints. For foundations that enclose below-grade spaces, the 2009 IRC adds requirements for the filling and sealing of the panel joints with approved materials conforming to the referenced standards. The code also clarifies that dampproofing is required using the same materials approved for cast-in-place concrete foundations.

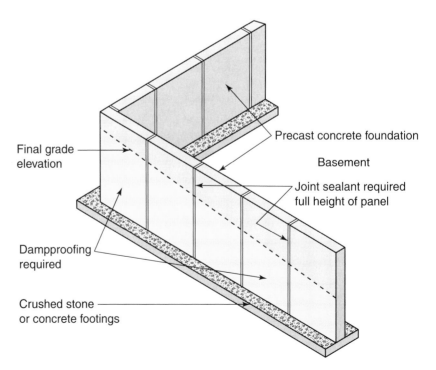

Precast concrete foundation system dampproofing

CHANGE TYPE: Clarification

CHANGE SUMMARY: Steel columns must be fabricated of not less than 3-inch-diameter Schedule 40 pipe.

2009 CODE: R407.3 Structural Requirements.   The columns shall be restrained to prevent lateral displacement at the bottom end. Wood columns shall not be less in nominal size than 4 inches by 4 inches (102 mm by 102 mm) ~~and~~. Steel columns shall not be less than 3-inch-diameter (76 mm) ~~standard pipe~~ Schedule 40 pipe manufactured in accordance with ASTM A53 Grade B or approved equivalent.

CHANGE SIGNIFICANCE: Nominal Pipe Size (NPS) is a North American set of standard sizes for pipes. Originally only a small selection of wall thicknesses were in use and were designated by weight class: standard weight (STD), extra-strong (XS), and double extra-strong (XXS). Pipe sizes are now specified with two numbers: the NPS in inches, and a schedule number. The older designation "3-inch-diameter standard pipe" corresponds to 3-inch Schedule 40 pipe. The previous reference to standard pipe for steel columns was intended to refer to ASTM A53 Grade B pipe with a 3-inch inside diameter, a Schedule 40 wall thickness of 0.211 inch, and an outside diameter of 3½ inches. Over time the reference to standard pipe has become unfamiliar to inspectors, resulting in a perceived inconsistent application of the requirement. The reference to Schedule 40 pipe manufactured in accordance with ASTM A 53 Grade B clearly defines the wall thickness and strength properties required for steel pipe columns. The code still recognizes other steel columns that provide equivalent performance characteristics.

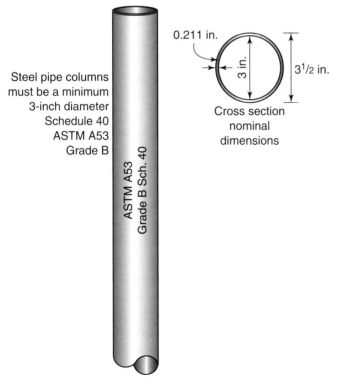

Steel pipe column

R407.3
Steel Columns

R408.1 and R408.2

Underfloor Space Ventilation

CHANGE TYPE: Modification

CHANGE SUMMARY: This change reestablishes a provision found in the 2003 IRC for reducing the required net area of ventilation openings to $\frac{1}{1500}$ of the underfloor area where the ground is covered with a vapor retarder.

2009 CODE: R408.1 Ventilation. The underfloor space between the bottom of the floor joists and the earth under any building (except space occupied by a basement) shall have ventilation openings through foundation walls or exterior walls. The minimum net area of ventilation openings shall not be less than 1 square foot (0.0929 m^2) for each 150 square feet (14 m^2) of underfloor space area, <u>unless the ground surface is covered by a Class 1 vapor retarder material. When a Class 1 vapor retarder material is used, the minimum net area of ventilation openings shall not be less than 1 square foot (0.0929 m^2) for each 1,500 square feet (140 m^2) of underfloor space area.</u> One such ventilating opening shall be within 3 feet (914 mm) of each corner of the building.

R408.2 Openings for Underfloor Ventilation. The minimum net area of ventilation openings shall not be less than 1 square foot (0.0929 m^2) for each 150 square feet (14 m^2) of underfloor area. One ventilation opening shall be within 3 feet (914 mm) of each corner of the building. Ventilation openings shall be covered for their height

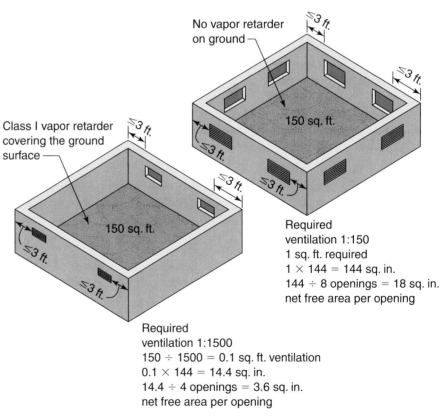

No vapor retarder on ground

≤3 ft.

≤3 ft.

150 sq. ft.

Required
ventilation 1:150
1 sq. ft. required
1 × 144 = 144 sq. in.
144 ÷ 8 openings = 18 sq. in.
net free area per opening

Class I vapor retarder covering the ground surface

≤3 ft.

150 sq. ft.

≤3 ft.

Required
ventilation 1:1500
150 ÷ 1500 = 0.1 sq. ft. ventilation
0.1 × 144 = 14.4 sq. in.
14.4 ÷ 4 openings = 3.6 sq. in.
net free area per opening

Crawl space ventilation

and width with any of the following materials, provided that the least dimension of the covering shall not exceed ¼ inch (6.4 mm)

1. Perforated sheet metal plates not less than 0.070 inch (1.8 mm) thick

2. Expanded sheet metal plates not less than 0.047 inch (1.2 mm) thick

3. Cast-iron grill or grating

4. Extruded load-bearing brick vents

5. Hardware cloth of 0.035 inch (0.89 mm) wire or heavier

6. Corrosion-resistant wire mesh, with the least dimension being $\frac{1}{8}$ inch (3.2 mm) thick.

Exception: The total area of ventilation openings shall be permitted to be reduced to $\frac{1}{500}$ of the underfloor area where the ground surface is covered with an approved Class I vapor retarder material and the required openings are placed so as to provide cross-ventilation of the space. The installation of operable louvers shall not be prohibited.

CHANGE SIGNIFICANCE: The 2006 IRC deleted a provision for reduced ventilation of crawl spaces where a vapor retarder covered the ground of the crawl space. This exception from the 2003 IRC is reinstated in the 2009 edition and is consistent with the similar provisions in the IBC. With the installation of the vapor retarder and the placement of openings to provide cross-ventilation, the required net area of the openings is only 10 percent of what is otherwise required or $\frac{1}{1500}$ of the crawl space area compared with $\frac{1}{150}$ of the area. The method using a vapor retarder is often considered the better option because of the potential for an uncovered ground surface to contribute significant amounts of moisture into the crawl space area. One difference from the previous language is that the code now specifies a Class I vapor retarder. A vapor retarder by definition is a membrane having a permeance rating of 1 perm or less. A Class I vapor retarder has a permeance rating of 0.1 perm or less and is therefore also considered vapor impermeable. Polyethylene sheeting is the most commonly used material to satisfy the requirement for a Class I vapor retarder.

R502.2.2.1 and Table R502.2.2.1

Deck Ledger Connection

CHANGE TYPE: Addition

CHANGE SUMMARY: Prescriptive methods for securely attaching a wood deck to the dwelling structure are now included in the IRC.

2009 CODE: <u>R502.2.2.1 Deck Ledger Connection to Band Joist.</u> <u>For decks supporting a total design load of 50 pounds per square foot (2394 Pa) [40 pounds per square foot (1915 Pa) live load plus 10 pounds per square foot (479 Pa) dead load], the connection between a deck ledger of pressure-preservative-treated Southern Pine, incised pressure-preservative-treated Hem-Fir or approved decay-resistant species, and a 2-inch (51 mm) nominal lumber band joist bearing on a sill plate or wall plate shall be constructed with ½-inch (12.7 m) lag screws or bolts with washers in accordance with Table R502.2.2.1. Lag screws, bolts and washers shall be hot-dipped galvanized or stainless steel.</u>

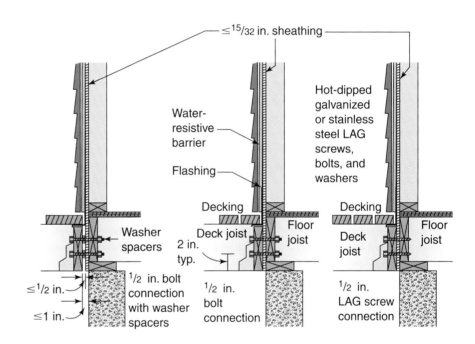

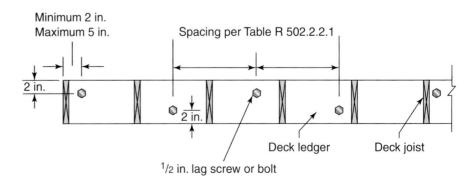

Deck ledger connection

TABLE R502.2.2.1 Fastener Spacing for a Southern Pine or Hem-Fir Deck Ledger and a 2-Inch Nominal Solid-Sawn Spruce-Pine-Fir Band Joist[c,f,g]
(Deck Live Load = 40 psf, Deck Dead Load = 10 psf)

Joist Span	6'-0" and Less	6'-1" to 8'-0"	8'-1" to 10'-0"	10'-1" to 12'-0"	12'-1" to 14'-0"	14'-1" to 16'-0"	16'-1" to 18'-0"
Connection Details	**On-Center Spacing of Fasteners[d, e]**						
½"-diameter lag screw with 15/32" maximum sheathing[a]	30	23	18	15	13	11	10
½"-diameter bolt with 15/32" maximum sheathing	36	36	34	29	24	21	19
½"-diameter bolt with 15/32" maximum sheathing and ½" stacked washers[b, h]	36	36	29	24	21	18	16

For SI: 1 inch = 25.4, 1 foot = 304.8 mm. 1 pound per square foot = 0.0479 kPa.
a. The tip of the lag screw shall fully extend beyond the inside face of the band joist.
b. The maximum gap between the face of the ledger board and face of the wall sheathing shall be ½."
c. Ledgers shall be flashed to prevent water from contacting the house band joist.
d. Lag screws and bolts shall be staggered in accordance with Section R502.2.2.1.1.
e. Deck ledger shall be minimum 2 × 8 pressure-preservative-treated No. 2 grade lumber or other approved materials as established by standard engineering practice.
f. When solid-sawn pressure-preservative-treated deck ledgers are attached to a minimum 1 inch thick engineered wood product (structural composite lumber, laminated veneer lumber, or wood structural panel band joist), the ledger attachment shall be designed in accordance with accepted engineering practice.
g. A minimum 1 by 9½ Douglas Fir laminated veneer lumber rimboard shall be permitted in lieu of the 2-inch nominal band joist.
h. Wood structural panel sheathing, gypsum board sheathing or foam sheathing not exceeding 1 inch in thickness shall be permitted. The maximum distance between the face of the ledger board and the face of the band joist shall be 1 inch.

R502.2.2.1.1 Placement of Lag Screws or Bolts in Deck Ledgers. The lag screws or bolts shall be placed 2 inches (51 mm) in from the bottom or top of the deck ledgers and between 2 and 5 inches (51 and 127 mm) in from the ends. The lag screws or bolts shall be staggered from the top to the bottom along the horizontal run of the deck ledger.

R502.2.2.2 Alternate Deck Ledger Connections. Deck ledger connections not conforming to Table R502.2.2.1 shall be designed in accordance with accepted engineering practice. Girders supporting deck joists shall not be supported on deck ledgers or band joists. Deck ledgers shall not be supported on stone or masonry veneer.

R502.2.2.3 Deck Lateral Load Connection. The lateral load connection required by Section R502.2.2 shall be permitted to be in accordance with Figure R502.2.2.3. Hold-down tension devices shall be provided in not less than two locations per deck, and each device shall have an allowable stress design capacity of not less than 1500 pounds (6672 N).

CHANGE SIGNIFICANCE: In the 2006 IRC, floor construction must be capable of accommodating all loads, and the deck provisions of Section R502.2.2 require positive connection to the primary structure to resist both vertical and lateral loads. Other than those performance requirements and the mention that nails subject to withdrawal were not satisfactory for the connection, the 2006 IRC contained no

R502.2.2.1 and Table R502.2.2.1 continues

R502.2.2.1 and Table R502.2.2.1 continued

specific methods for attaching a deck to the structure. Many have expressed concern that the traditional methods of deck construction in use across the country for many years, typically approved locally as conventional construction without an engineered design, may not be adequate. Deck failures have occurred occasionally due to improper attachment. The new prescriptive methods for deck ledger connection to the band joist were developed through engineering analysis and testing of the various materials listed and intend to provide an adequate in-service safety factor. The intent of these prescriptive provisions is to provide guidance to designers, contractors, and building officials in determining proper and safe attachment of the deck to the structure and to ensure consistent application of the code.

The prescribed methods of attachment permit fastening to a 2-inch nominal solid-sawn lumber band joist or a minimum 1-inch by 9½-inch Douglas fir laminated veneer lumber (LVL) rimboard. Attachment to other engineered wood products, such as structural composite lumber or wood structural panel band joists, requires a design in accordance with accepted engineering practice. For bolt connections, the code permits a maximum distance of 1 inch between the face of the ledger board and the face of the band joist. Bolt and lag screw spacing is otherwise based on a maximum sheathing thickness of $^{15}/_{32}$ inch. In addition to the maximum spacing requirements, the hot-dipped galvanized or stainless steel lag screws or bolts must be located 2 inches from the top or bottom of the deck ledger and from 2 to 5 inches from the end of the ledger.

Section R502.2.2 requires a connection designed for both vertical and lateral loads. Hold-down tension devices are permitted as a prescriptive option to satisfy the lateral load requirement when they are installed on at least two locations per deck and each device has an allowable stress design capacity of not less than 1500 pounds. The detail for hold-down devices as shown in the figure is not a code requirement but if used will comply with Section R502.2.2 even in the most severe Seismic Design Categories covered by the IRC.

CHANGE TYPE:　Clarification

CHANGE SUMMARY:　New text clarifies that installation of engineered wood products including lateral support to prevent rotation is determined by the installation instructions of the manufacturer.

2009 CODE:　**R502.7 Lateral Restraint at Supports.**　Joists shall be supported laterally at the ends by full-depth solid blocking not less than 2 inches (51 mm) nominal in thickness; or by attachment to a full-depth header, band or rim joist, or to an adjoining stud, or shall be otherwise provided with lateral support to prevent rotation.

Exceptions:

1. Trusses, structural composite lumber, structural glued-laminated members and I-joists shall be supported laterally as required by the manufacturer's recommendations.

2. In Seismic Design Categories D_0, D_1 and D_2, lateral restraint shall also be provided at each intermediate support.

R502.7.1 Bridging.　Joists exceeding a nominal 2 inches by 12 inches (51 mm by 305 mm) shall be supported laterally by solid blocking, diagonal bridging (wood or metal), or a continuous 1-inch-by-3-inch (25.4 mm by 76 mm) strip nailed across the bottom of joists perpendicular to joists at intervals not exceeding 8 feet (2438 mm).

R502.7

Lateral Restraint for Wood Joists

Lateral restraint for engineered wood joists and trusses at bearing points

R502.7 continues

R502.7 continued

Exception: Trusses, structural composite lumber, structural glued-laminated members and I-joists shall be supported laterally as required by the manufacturer's recommendations.

CHANGE SIGNIFICANCE: The prescribed lateral support requirements of Section R502.7 are intended to apply only to solid-sawn lumber joists. The new text clarifies that the requirements do not apply to engineered wood products such as plate-connected trusses, I-joists, glued-laminated lumber, and structural composite lumber. Engineered wood products may require other means of lateral restraint, and the recommendations of the manufacturer will govern those installations.

CHANGE TYPE: Modification

CHANGE SUMMARY: The prescriptive provisions of cold-formed steel framing now apply to three-story buildings, an increase from the previous limitation of two stories. This section has been reorganized and modified to clarify the application of the code. New provisions, tables, and figures provide more options for the design and construction of dwellings using the prescriptive cold-formed steel framing provisions.

2009 CODE: R505.1.1 Applicability Limits. The provisions of this section shall control the construction of cold-formed steel floor framing for buildings not greater than 60 feet (18 288 mm) in length perpendicular to the joist span, not greater than 40 feet (12 192 mm) in

R505
Cold-Formed Steel Floor Framing

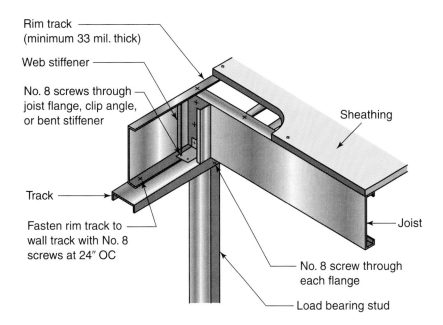

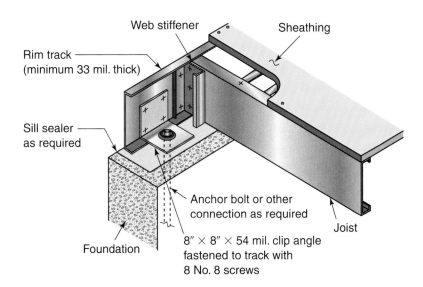

Cold-formed steel floor framing details

R505 continues

R505 continued   width parallel to the joist span, and ~~not greater~~ less than ~~two~~ or equal to three stories ~~in height~~ above grade plane. Cold-formed steel floor framing constructed in accordance with the provisions of this section shall be limited to sites subjected to a maximum design wind speed of 110 miles per hour (49 m/s), Exposure B or C, and a maximum ground snow load of 70 pounds per square foot (3.35 kPa).

R505.1.2 In-Line Framing.   When supported by cold-formed steel framed walls in accordance with Section R603, cold-formed steel floor framing shall be constructed with floor joists located ~~directly~~ in line with load-bearing studs located below the joists in accordance with Figure R505.1.2 and the tolerances specified as follows:

1. ~~with a~~ The maximum tolerance shall be ~~of~~ ¾ inch (19.1 mm) between the centerline of the horizontal framing member and the centerline of the vertical framing member. ~~between the center lines of the joist and the stud~~

2. Where the centerline of the horizontal framing member and bearing stiffener are located to one side of the centerline of the vertical framing member, the maximum tolerance shall be $\frac{1}{8}$ inch (3 mm) between the web of the horizontal framing member and the edge of the vertical framing member.

R505.1.3 Floor Trusses.   ~~The design, quality assurance, installation and testing of~~ Cold-formed steel trusses shall be designed, braced and installed in accordance with ~~AISI Standard for Cold-formed Steel Framing-Truss Design (COFS/Truss)~~ AISI S100, Section D4. Truss members shall not be notched, cut or altered in any manner without an approved design.

R505.3.1 Floor to Foundation or Load-bearing Wall Connections.   Cold-formed steel framed floors shall be anchored to foundations, wood sills or load-bearing walls in accordance with Table R505.3.1(1) and Figure R505.3.1(1), R505.3.1(2), R505.3.1(3), R505.3.1(4), R505.3.1(5) or R505.3.1(6). Anchor bolts shall be located not more than 12 inches (305 mm) from corners or the termination of bottom tracks. Continuous cold-formed steel joists supported by interior load-bearing walls shall be constructed in accordance with Figure R505.3.1(7). Lapped cold-formed steel joists shall be constructed in accordance with Figure R505.3.1(8). End floor joists constructed on foundation walls parallel to the joist span shall be doubled unless a C-shaped bearing stiffener, sized in accordance with Section R505.3.4, is installed web-to-web with the floor joist beneath each supported wall stud, as shown in Figure R505.3.1(9). Fastening of cold-formed steel joists to other framing members shall be in accordance with Section R505.2.4 and Table R505.3.1(2).

Because this code change deleted, added, and modified substantial portions of Section R505, the entire code change text is too extensive to be included here. Refer to Code Change RB127-07/08 in the

2009 IRC Code Changes Resource Collection for the complete text and history of the code change.

CHANGE SIGNIFICANCE: Section R505 has undergone significant revision and updating to incorporate provisions of the new American Iron and Steel Institute (AISI) *Standard for Cold-Formed Steel Framing—Prescriptive Method for One- and Two-Family Dwellings* (AISI S230-2007). The most notable change to the prescriptive methods is in the scope of application, where the height limitation has increased from two-story to three-story buildings, consistent with the height limits of conventional wood frame construction. Terminology has been updated throughout the text to reflect current usage by industry and to provide consistency in the code. For example, "Reference Gage Number" has been deleted, because gage is no longer used by industry in referencing structural members, and "uncoated steel thickness" has been changed to "base steel thickness." Many minor modifications improve the organization, clarity, and usability of the code provisions. In general, the revisions provide more options for the design and construction of dwellings using the prescriptive cold-formed steel framing provisions.

The tolerances for floor joists located in line with cold-formed steel studs have been revised to account for the special case of the bearing stiffener located on the back side of the joist. Provisions concerning web holes and web hole adjustments have been modified and placed into one location. The code user now has the choice to reinforce nonconforming holes, patch nonconforming holes, or design nonconforming holes in accordance with accepted engineering practice. Provisions for joist bracing and blocking have been divided into four distinct sections: joist top flange bracing, joist bottom flange bracing and blocking, blocking at interior bearing supports, and blocking at cantilevers. Four tables have been added detailing the design of clip angle bearing stiffeners in order to permit more options for the builder. Many of the new illustrations relate to the location or fastening requirements for the installation of blocking and bearing stiffeners. Separate joist header and trimmer details are provided for 6-foot-wide and 8-foot-wide openings.

Table R602.3(1)

Fastener Schedule for Structural Members

CHANGE TYPE: Modification

CHANGE SUMMARY: Table R602.3(1) has been reorganized and updated to reflect currently accepted industry standards and manufacturer's recommendations. The fastening requirements for ceiling joist and rafter tie connections to rafters have been deleted because these connection requirements appear in Table R802.5.1.9, Rafter/Ceiling Joist Heel Joint Connections.

2009 CODE:

TABLE R602.3(1)   Fastener Schedule for Structural Members

Item	Description of Building Elements	Number and Type of Fastener[a,b,c]	Spacing of Fasteners
	Roof		
1	Blocking between joists or rafters to top plate, toe nail	3-8d (2½″ × 0.113″)	—
2	Ceiling joists to plate <u>or girder/beam</u>, toe nail	3-8d (2½″ × 0.113″)	—
3	Ceiling joist <u>not attached to parallel rafter</u>, laps over partitions, face nail	3-10d (3″ × 0.128″)	—
	~~Ceiling joist to parallel rafters, face nail~~	~~3-10d (3″ × 0.128″)~~	—
4	Collar tie to rafter, face nail, or 1¼″ × 20-gauge ridge strap	3-10d (3″ × 0.128″)	—
5	Rafter to plate <u>or girder/beam</u>, toe nail	2-16d (3½″ × 0.135″)	—
6	Roof rafters to ridge, valley, or hip rafters:		
	toe nail	4-16d (3½″ × 0.135″)	—
	face nail	3-16d (3½″ × 0.135″)	—
	~~Rafter ties to rafters, face nail~~	~~3-8d (2 ½″ × 0.113″)~~	—
	Wall		
7	Built-up corner studs	10d (3″ × 0.128″)	24″ O.C.
8	Built-up header, two pieces with ½″ spacer	16d (3½″ × 0.135″)	16″ O.C. along each edge
9	Continued header, two pieces	16d (3½″ × 0.135″)	16″ O.C. along each edge
10	Continuous header to stud, toe nail	4-8d (2½″ × 0.113″)	—
11	Double studs, face nail	10d (3″ × 0.128″)	24″ O.C.
12	Double top plates, face nail	10d (3″ × 0.128″)	24″ O.C.
13	Double top plates, minimum 24-in. offset of end joints, face nail in lapped area	8-16d (3½″ × 0.135″)	—
14	Sole plate to joist or blocking, face nail	16d (3½″ × 0.135″)	16″ O.C.
15	Sole plate to joist or blocking at braced wall panels	3-16d (3½″ [0.135″)	16″ O.C.
16	Stud to sole plate, toe nail	3-8d (2½″ × 0.113″) or 2-16d (3½″ × 0.135″)	—
17	Top or sole plate to stud, end nail	2-16d (3½″ × 0.135″)	—
18	Top plates, laps at corners and intersections, face nail	2-10d (3″ × 0.128″)	—
19	1″ brace to each stud and plate, face nail	2-8d (2½″ × 0.113″) 2 staples, 1¾″	— —
20	1″ × 6″ sheathing to each bearing, face nail	2-8d (2½″ × 0.113″) 2 staples, 1¾″	—
21	1″ × 8″ sheathing to each bearing, face nail	2-8d (2½″ × 0.113″) 3 staples, 1¾″	—
22	Wider than 1″ × 8″ sheathing to each bearing, face nail	3-8d (2½″ × 0.113″) 4 staples, 1¾″	—

Floor

23	Joist to sill or girder, toe nail	3-8d (2½″ × 0.113″)	—
24	1″ × 6″ subfloor or less to each joist, face nail	2-8d (2½″ × 0.113″) 2 staples 1¾″	—
25	2″ subfloor to joist or girder, blind and face nail	2-16d (2½″ × 0.135″)	—
26	Rim joist to top plate, toe nail (roof applications also)	8d (2½″ × 0.113″) 6″ o.c.	6″ O.C.
27	2″ planks (plank & beam—floor & roof)	2-16d (2½″ × 0.135″)	At each bearing

Beams and Girders

28	Built-up girders and beams, 2-in. lumber layers	10d (3″ × 0.128″)	Nail each layer as follows: 32″ O.C. at top and bottom and staggered. Two nails at ends and at each splice.
29	Ledger strip supporting joists or rafters	3-16d (2½″ × 0.135″)	At each joist or rafter

			Spacing of Fasteners	
Item	Description of Building Materials	Description of Fastener[b,c,e]	Edges (in.)[i]	Intermediate Supports (in.)[c,e]
Wood structural panels, subfloor, roof and _interior_ wall sheathing to framing, and particleboard wall sheathing to framing				
30	~~5/16~~ ³⁄₈″-½″	6d common (2″ × 0.113″) nail (subfloor, wall)[j] 8d common (2½″ × 0.131″) nail (roof)[f]	6	12[g]
31	19/32″–1″	8d common nail (2½″ × 0.131″)	6	12
32	1⅛″ -1¼″	10d common (3″ × 0.148″) nail or 8d (2½″ × 0.120″) deformed nail	6	12
Other wall sheathing[h]				
33	½″ structural cellulosic fiberboard sheathing	1½″ galvanized roofing nail; ~~8d common (2-½″ × 0.131″) nail;~~ ⁷⁄₁₆″ crown or 1″ crown staple 16 ga., ~~1 ½~~ 1¼″ long	3	6
34	²⁵⁄₃₂″ structural cellulosic fiberboard sheathing	1¾″ galvanized roofing nail; ~~8d common (2-½″ × 0.131″) nail;~~ ⁷⁄₁₆″ crown or 1″ crown staple 16 ga., ~~1 ¾~~ 1½″ long	3	6
35	½″ gypsum sheathing[d]	1½″ galvanized roofing nail; ~~6d common (2″ × 0.113″) nail;~~ staple galvanized, 1½″ long; 1¼″ screws, Type W or S	~~4~~ 7	~~8~~ 7
36	⅝″ gypsum sheathing[d]	1¾″ galvanized roofing nail; ~~8d common (2-1/2″ × 0.131″) nail;~~ staple galvanized, 1⅝″ long; 1⅝″ screws, Type W or S	~~4~~ 7	~~8~~ 7

Portions of table not shown. (No change to current text.)

Notes a. through g. (No change to current text.)

h. Gypsum sheathing shall conform to ASTM ~~C 79~~ C 1396 and shall be installed in accordance with GA 253. Fiberboard sheathing shall conform to ASTM C 208.

i. (No change to current text.)

j. For regions having a basic wind speed of 85 mph or greater, the nail size and attachment schedule of Table 602.3(3) shall be used for attaching wood structural panel wall sheathing.

R602.3(1) continues

R602.3(1) continued

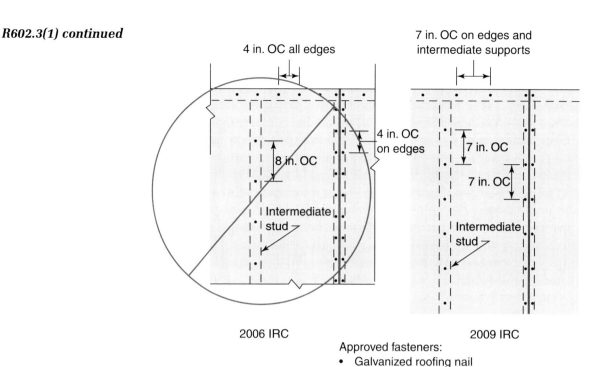

Gypsum board wall sheathing fastenening

CHANGE SIGNIFICANCE: The fastening requirements for solid-sawn lumber framing members in Table R602.3(1) have been reorganized into four categories related to roof, wall, floor, and beam/girder locations, and each condition has been given an item number. The new format makes it easier to locate the appropriate nailing requirement.

Prior to the 2006 IRC, Section R802.3.1, Ceiling Joist and Rafter Connections, referenced the fastening requirements of Table R602.3(1) and Table R802.5.1.9, Rafter/Ceiling Joist Heel Joint Connections, for ceiling joist to rafter nailing necessary to resist the outward thrust of the rafters. Only Table R602.3(1) was referenced for the nailing requirements at the laps of ceiling joists and the connection of rafter ties to the rafter. In the 2006 edition, the IRC referenced Table R802.3.1 for all three conditions. Though the reference to the generic fastener table, Table R602.3(1), was deleted from Section R802.3.1, the differing nailing requirements for these conditions remained in Table R602.3(1), causing some confusion as to the proper fastening requirements. For conditions where the framing member is fastened to parallel rafters, this change removes the applicable items from Table R602.3(1): the face nailing of ceiling joists where they lap over partitions, ceiling joist to parallel rafters, and rafter ties to rafters. The correct values appear in the appropriate referenced Table R802.5.1.9. Note that Table R602.3(1) still applies to ceiling joist lap nailing when the joists are not attached to parallel rafters.

The sheathing fastener schedule has been updated to reflect currently accepted industry recommendations and commonly used or available materials. Common nails are not recommended for attaching gypsum sheathing due to their relatively smaller diameter head and head thickness. Consequently, the heads of common nails often tear the face paper of the gypsum sheathing, and common nails have therefore been deleted from the approved gypsum sheathing attachment methods. The prescribed fastener spacing at the edges and in the field of gypsum sheathing panels is now 7 inches and matches the attachment requirements for gypsum board used as wall bracing. This is an increase from 4-inch spacing at the edges and a decrease from 8-inch spacing at intermediate supports as required in the 2006 IRC. The change to footnote *h* replaces the reference to an obsolete standard with the current ASTM C 1396 *Standard Specification for Gypsum Board*.

Wood structural panels with a thickness of $\frac{5}{16}$ inch are no longer commonly available or used in construction. The minimum thickness of wood structural panels recognized in Table R602.3(1) is now $\frac{3}{8}$ inch. The heading for wood structural panel sheathing clarifies that Table R602.3(1) does not apply to exterior wood structural panel wall sheathing. The fastening requirements of Table R602.3(3) now apply to such exterior wall sheathing and are based on wind speed, wind exposure category, and thickness of the sheathing material.

The changes to the fastening requirements for fiberboard sheathing reflect the appropriate staple crown size for providing adequate shear capacity of the sheathing. Due to their relatively small heads, common nails are no longer approved for attachment of fiberboard sheathing. Minimum staple leg lengths are shortened based on available proprietary ASTM E 72 tests and fastener withdrawal calculations.

R602.3 and Table R602.3(3)

Wood Structural Panel Wall Sheathing Used to Resist Wind Pressures

CHANGE TYPE: Modification

CHANGE SUMMARY: The component and cladding wind load requirements of Section R301.2.1 and Table R301.2(2) are now referenced in Section R602.3. Wood structural panels used as exterior wall sheathing must comply with the new Table R602.3(3), which now establishes minimum requirements for fastening, panel thickness, span ratings, and stud spacing based on design wind speed and wind exposure category.

2009 CODE: R602.3 Design and Construction. Exterior walls of wood frame construction shall be designed and constructed in accordance with the provisions of this chapter and Figures R602.3(1)

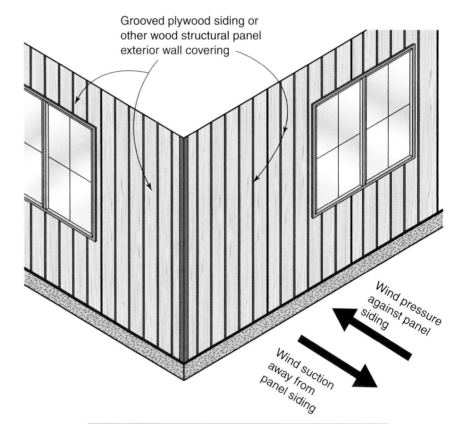

Siding Panel Thickness	Nail Size and Spacing	Stud Spacing	Maximum Wind Speed Wind Exposure Category		
			A	B	C
7/16 in.	8^d common 6 in. OC edges 12 in. OC field	16 in. OC	130 MPH	110 MPH	105 MPH

Wood structural panel siding to resist wind pressures

and R602.3.(2) or in accordance with AF&PA's NDS. Components of exterior walls shall be fastened in accordance with Tables R602.3(1) through R602.3(4). ~~Exterior walls covered with foam plastic sheathing shall be braced in accordance with Section R602.10.~~ Structural <u>wall</u> sheathing shall be fastened directly to structural framing members. <u>Exterior wall coverings shall be capable of resisting the appropriate wind pressures listed in Table R301.2(2) adjusted for height and exposure using Table R301.2(3). Wood structural panel sheathing used for exterior walls shall conform to the requirements of Table R602.3(3).</u>

~~TABLE R602.3(3)~~   ~~Wood Structural Panel Wall Sheathing~~

~~Panel Span Rating~~	~~Panel Nominal Thickness (inch)~~	~~Maximum Stud Spacing (inches)~~	
		~~Siding nailed to:[a]~~	
		~~Stud~~	~~Sheathing~~
~~12/0, 16/0, 20/0, or wall × 16 o.c.~~	~~5/16, 3/8~~	~~16~~	~~16[b]~~
~~24/0, 24/16, × 24 o.c.~~	~~3/8, 7/10, 15/32, 1/2~~	~~24~~	~~24[c]~~

~~For SI: 1 inch = 25.4 mm.~~
~~a. Blocking of horizontal joints shall not be required.~~
~~b. Plywood sheathing 3/8-inch thick or less shall be applied with long dimension across studs.~~
~~c. Three-ply plywood panels shall be applied with long dimension across studs.~~

TABLE R602.3(3)   Requirements for Wood Structural Panel Wall Sheathing Used to Resist Wind Pressures[a,b,c]

Minimum Nail		Minimum Wood Structural Panel Span Rating	Minimum Nominal Panel Thickness (inches)	Maximum Wall Stud Spacing (inches)	Panel Nail Spacing		Maximum Basic Wind Speed (mph)		
					Edges (inches o.c.)	Field (inches o.c.)	Wind Exposure Category		
Size	Penetration (inches)						B	C	D
6d Common (0.113" x 2.0")	1.5	24/0	3/8	16	6	12	110	90	85
8d Common (0.131" x 2.5")	1.75	24/16	7/16	16	6	12	130	110	105
				24	6	12	110	90	85

a. Panel strength axis parallel or perpendicular to supports. Three-ply plywood sheathing with studs spaced more than 16 inches on center shall be applied with panel strength axis perpendicular to supports.
b. Table is based on wind pressures acting toward and away from building surfaces per R301.2. Lateral bracing requirements shall be in accordance with R602.10.
c. Wood structural panels with span ratings of Wall-16 or Wall-24 shall be permitted as an alternate to panels with a 24/0 span rating. Plywood siding rated 16 oc or 24 oc shall be permitted as an alternate to panels with a 24/16 span rating. Wall-16 and plywood siding 16 oc shall be used with studs spaced a maximum of 16 inches on center.

R602.3 and Table R602.3(3) continues

R602.3 and Table R602.3(3)
continued

CHANGE SIGNIFICANCE: Previously, Table R602.3(3) provided the minimum thickness of wood structural panels attached to studs 16 inches and 24 inches on center based on the panel floor and roof span rating. Wood structural panel fastening requirements were located in Table R602.3(1). The new Table R602.3(3) is the result of engineering analysis on the effects of wind pressures acting toward and away from building surfaces to meet the requirements of Section R301.2.1. The analysis considered panel bending, stiffness, nail withdrawal, and nail head pull-through. Table R602.3(3) now contains minimum fastening requirements for exterior wood structural panel wall sheathing, including panel thickness, span ratings, and stud spacing based on design wind speed and wind exposure category. In most cases, the change will not have a significant impact on construction practices. The most common application of wood structural panel sheathing is $^{7}/_{16}$-inch-thick panels attached to studs 16 inches on center. This application is satisfactory for wind speeds up to 110 mph except in Exposure Category D (shorelines of open water areas such as the Great Lakes and the West coast). This application does require a greater nail size than was previously required in Table R602.3.1—8d common compared with 6d common.

Because the charging language of Section R602.10 applies to all exterior walls regardless of the exterior wall covering type, the reference to the wall bracing section for foam plastic sheathing was considered unnecessary and has been deleted from Section R602.3.

CHANGE TYPE: Modification

CHANGE SUMMARY: A habitable attic, a new defined term in the 2009 IRC, is treated the same as a typical roof and ceiling forming an attic in determining wood stud size and spacing in Table R602.3(5).

2009 CODE:

CHANGE SIGNIFICANCE: A habitable attic is an enclosed space between the ceiling of the top story and the roof assembly. Whether finished or unfinished, such a space meeting the minimum room area

Table R602.3(5)
Size, Height, and Spacing of Wood Studs

TABLE R602.3(5) Size, Height, and Spacing of Wood Studs[a]

		Bearing Walls				Nonbearing Walls	
Stud Size (inches)	Laterally Unsupported Stud Height[a] (feet)	Maximum Spacing When Supporting ~~roof and ceiling~~ a Roof-Ceiling Assembly or a Habitable Attic Assembly Only (inches)	Maximum Spacing When Supporting One Floor ~~roof and ceiling~~ plus a Roof-Ceiling Assembly or a Habitable Attic Assembly (inches)	Maximum Spacing When Supporting Two Floors ~~roof and ceiling~~ plus a Roof-Ceiling Assembly or a Habitable Attic Assembly (inches)	Maximum Spacing When Supporting One Floor Only (inches)	Laterally Unsupported Stud Height[a] (feet)	Maximum Spacing (inches)
2 × 3[b]	—	—	—	—	—	10	16
2 × 4	10	24[c]	16[c]	—	24	14	24
3 × 4	10	24	24	16	24	14	24
2 × 5	10	24	24	—	24	16	24
2 × 6	10	24	24	16	24	20	24

For SI: 1 inch = 25.4 mm, 1 foot = 304.8 mm.
a. Listed heights are distances between points of lateral support placed perpendicular to the plane of the wall. Increases in unsupported height are permitted where justified by analysis.
b. Shall not be used in exterior walls.
c. A habitable attic assembly supported by 2 × 4 studs is limited to a roof span of 32 feet. Where the roof span exceeds 32 feet, the wall studs shall be increased to 2 × 6, or the studs shall be designed in accordance with accepted engineering practice.

Table R602.3(5) continues

Table R602.3(5) continued   and ceiling height requirements for habitable spaces and having a minimum floor live load of 30 psf is considered a habitable attic but not a story. Placement of habitable attics in the wood stud table clarifies that wood studs of a size, height, and spacing adequate for carrying a roof and ceiling also are adequate for supporting a habitable attic. Footnote *d* places a limitation of 32 feet for the roof span when using 2 × 4 studs to support a habitable attic. For greater roof spans, the code requires not less than 2 × 6 studs or an engineered design.

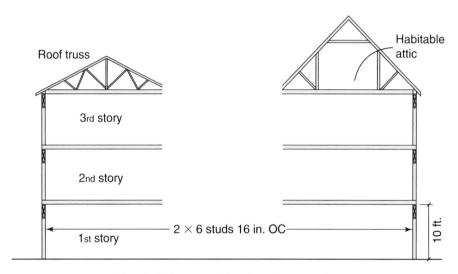

Wood studs supporting two floors and a roof-ceiling assembly or a habitable attic

CHANGE TYPE: Modification

CHANGE SUMMARY: When a metal tie is required across the opening of a notched or drilled top plate, the tie must now extend at least 6 inches beyond each side of the opening. To reduce the possibility of splitting the wood plate, the length of the nails used to attach the metal tie has been reduced from 3½ inches to 1½ inches. The nails must have a minimum diameter of 0.148 inch.

2009 CODE: **R602.6.1 Drilling and Notching of Top Plate.**   When piping or ductwork is placed in or partly in an exterior wall or interior load-bearing wall, necessitating cutting, drilling, or notching of the top plate by more than 50 percent of its width, a galvanized metal tie of not less than 0.054 inch thick (1.37 mm) (16 ga) and 1½ inches (38 mm) wide shall be fastened across and to the plate at each side of the opening with not less than eight ~~16d~~ 10d (0.148 inch diameter) nails having a minimum length of 1½ inches (38 mm) at each side or equivalent. The metal tie must extend a minimum of 6 inches past the opening. See Figure R602.6.1.

Exception: When the entire side of the wall with the notch or cut is covered by wood structural panel sheathing

R602.6.1
Drilling and Notching of Top Plate

Metal tie required at notched top plate

R602.6.1 continues

R602.6.1 continued

CHANGE SIGNIFICANCE: Drilling or notching of top plates is sometimes necessary for the installation of plumbing or mechanical piping or vents. When top plates are drilled or notched more than 50 percent of their width, the code requires a metal strap across the opening to tie the weakened plates together and resist tension in the direction of the plates (parallel to the wall). Previously, at least eight 16d nails were required to fasten the tie on each side of the opening, and the minimum length of the tie was not specified. The change to shorter nails and a tie of sufficient length to extend at least 6 inches beyond the opening on each side is in response to concerns of excessive splitting of the top plates. Nails with a 0.148 inch diameter and a 1½ inch length exceed the shear capacity of a 16d box nail, which is 0.135 inch diameter and 3½ inches long. Though a 10d nail is 3 inches long, the intent of this change is that 1½-inch-long nails with a diameter equivalent to 10d common nails (0.148 inch) provide adequate shear capacity and satisfy the requirement. Such nails are commonly available and typically used for joist hangers and other connectors.

CHANGE TYPE: Modification

CHANGE SUMMARY: The wood frame wall bracing provisions of Section R602.10 have been entirely rewritten to provide technical accuracy and clarity. The code no longer differentiates between exterior and interior braced wall lines. The terms *braced wall line* and *braced wall panel* are more precisely defined. New language clarifies how braced wall lines are measured and when mixing of bracing methods is permitted. The charging language more clearly prescribes the paths for compliance—intermittent bracing, continuous sheathing, or an engineered design.

2009 CODE: **R602.10 Wall Bracing.** ~~All exterior walls~~ Buildings shall be braced in accordance with this section. ~~In addition, interior braced wall lines shall be provided in accordance with Section R602.10.1.1. For buildings in Seismic Design Categories D0, D1 and D2, walls shall be constructed in accordance with the additional requirements of Sections R602.10.9, R602.10.11, and R602.11.~~ Where a building, or portion thereof, does not comply with one or more of the bracing requirements in this section, those portions shall be designed and constructed in accordance with Section R301.1.

> **Exception:** Detached one- and two-family dwellings located in Seismic Design Category C are exempt from the seismic bracing requirements of this section. Wind speed provisions for bracing shall be applicable to detached one- and two-family dwellings.

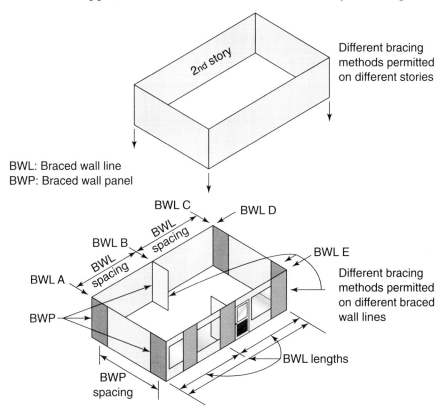

BWL: Braced wall line
BWP: Braced wall panel

Braced wall lines and braced wall panels

R602.10 continues

R602.10 continued

R602.10.1.1 Braced Wall Panels. Braced wall panels shall be constructed in accordance with the intermittent bracing methods specified in Section R602.10.2, or the continuous sheathing methods specified in Sections R602.10.4 and R602.10.5. Mixing of bracing method shall be permitted as follows:

1. Mixing bracing methods from story to story is permitted.

2. Mixing bracing methods from braced wall line to braced wall line within a story is permitted, except that continuous sheathing methods shall conform to the additional requirements of Sections R602.10.4 and R602.10.5.

3. Mixing bracing methods within a braced wall line is only permitted in Seismic Design Categories A and B, and detached dwellings in Seismic Design Category C. The length of required bracing for the braced wall line with mixed sheathing types shall have the higher bracing length requirement, in accordance with Tables R602.10.1.2(1) and R602.10.1.2(2), of all types of bracing used.

SECTION R202 DEFINITIONS

Braced Wall Line. A straight line through the building plan that represents the location of the lateral resistance provided by the wall bracing ~~series of braced wall panels in a single story constructed in accordance with Section R602.10 for wood framing or Section R603.7 or R301.1.1 for cold-formed steel framing to resist racking from seismic and wind forces.~~

Braced Wall Panel. A full-height section of ~~a braced wall line~~ wall constructed to resist in-plane shear loads through interaction of framing members, sheathing material, and anchors. The panel's length meets the requirements of its particular bracing method and contributes to the total amount of bracing required along its braced wall line in accordance with Section R602.10.1. ~~in accordance with Section R602.10 for wood framing or Section R603.7 or R301.1.1 for cold-formed steel framing, which extend the full height of the wall.~~

Because numerous code changes deleted, added, and modified substantial portions of Section R602.10, the entire code change text is too extensive to be included here. Refer to the *2009 IRC Code Changes Resource Collection* for the complete text and history of code changes related to Section R602.10.

CHANGE SIGNIFICANCE: The wall bracing provisions in Section R602.10 have undergone a major overhaul from the 2006 provisions, reorganizing and revising the text for technical accuracy and clarity, and intend to make the section more user friendly. Many of the changes are the result of work by the ICC Ad Hoc Committee on Wall Bracing including engineering analysis of the prescriptive methods used to resist lateral seismic and wind forces. Such analysis revealed that clarification was needed in the bracing requirements related to wind forces and led to placing the wind and seismic provisions in separate tables with the more restrictive requirements controlling the amount of bracing.

The definitions of *braced wall line* and *braced wall panel* are revised for clarity and uniform application. Previous definitions were circular in that they used the other term in the definition—a *braced wall line* was a series of braced wall panels and a *braced wall panel* was a segment of a braced wall line. *Braced wall panel* is now defined in engineering terms to clarify its purpose to resist in-plane shear loads. In the same way, *braced wall line* is now defined by its purpose—to designate the location of the lateral resistance system provided by the wall bracing. New text in Section R602.10.1 precisely describes where a braced wall line begins and ends in determining its length.

The code no longer differentiates between exterior and interior braced wall lines. The previous language caused confusion, leading some to believe that interior braced wall lines had to begin and end inside the building. In a structural sense, all braced wall lines act in the same way regardless where they are located on the building, and this change removes the unnecessary and sometimes confusing language.

Section R301.2.2 exempts one- and two-family dwellings located in Seismic Design Category (SDC) C from the seismic requirements of the code. By also inserting this exemption into the beginning of Section R602.10, the code clarifies the application of the wall bracing provisions for these dwellings. One- and two-family dwellings located in SDC C need only comply with the requirements for buildings located in SDCs A and B, which have no special seismic requirements. This new language also clarifies that there is no similar reduction in bracing requirements of one- and two-family dwellings for resisting wind forces.

By introducing the term *intermittent bracing* to define the use of isolated braced wall panels within a braced wall line, the code now clearly distinguishes the three separate paths for compliance with the bracing requirements—the prescriptive methods using either intermittent braced wall panels or continuous wall sheathing, or bracing (shear walls) in accordance with an engineered design.

Previously, the code was silent on mixing various types of bracing methods on the same building, though the practice of mixing methods was very common. The code now specifically allows mixing of methods between stories and from one wall line to the next. Different bracing methods are also permitted within the same wall line in SDCs A, B, and C only.

R602.10.1.2
Length of Wall Bracing

CHANGE TYPE: Modification

CHANGE SUMMARY: Lateral bracing requirements related to wind loads and seismic loads have been placed in separate tables. The greater tabular value from the two tables based on the building location applies. The total amount of bracing required is the product of all applicable adjustment factors. The amount of required bracing is now expressed as length in feet rather than as a percentage of the braced wall line. After all adjustments are made, the minimum total length of bracing in a braced wall line must be at least 48 inches. Walls perpendicular to the braced wall line do not count toward the bracing amount required in the direction of the braced wall line under consideration. Provisions were added for exterior braced wall panels that are subjected to wind uplift. Also, a footnote to the wind bracing table permits the required bracing length for methods other than let-in bracing in one-story or the top story of two or three-story buildings to be reduced when tie-down devices are provided at braced wall panels.

2009 CODE: <u>**R602.10.1.2 Length of Bracing.** The length of bracing along each braced wall line shall be the greater of that required by the design wind speed and braced wall line spacing in accordance with Table R602.10.1.2(1) as adjusted by the factors in the footnotes, or the Seismic Design Category and braced wall line length in accordance with Table R602.10.1.2(2) as adjusted by the factors in Table</u>

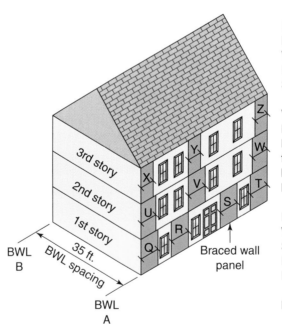

Example:
Building location
Wind = 90 MPH
Exposure category B
SDC = B

Wind controls:
Minimum bracing
length is determined
from Table R602.10.1(1)
based on braced wall
line spacing

Intermittent bracing method
WSP: Wood structural panel
SFB: Structural fiber board
PBS: Particle board
PCP: Portland cement plaster
OR
HPS: Hardboard panel siding

Minimum tabular values for length of wall bracing
1st story Q + R + S + T = 20.5 ft.*
2nd story U + V + W = 14 ft.*
3rd story X + Y + X = 7.5 ft.*

BWL: Braced wall line   *Subject to adjustment factors

Determining length of required wall bracing

R602.10.1.2(3). Only walls that are parallel to the braced wall line shall be counted toward the bracing requirement of that line, except angled walls shall be counted in accordance with Section R602.10.1.3. In no case shall the minimum total length of bracing in a braced wall line, after all adjustments have been taken, be less than 48 inches total.

Because numerous code changes deleted, added, and modified substantial portions of Section R602.10, the entire code change text is too extensive to be included here. Refer to the *2009 IRC Code Changes Resource Collection* for the complete text and history of code changes related to Section R602.10.

CHANGE SIGNIFICANCE: As part of the reorganization and technical modification of the wall bracing provisions, the bracing requirements to resist wind and seismic forces are placed in separate tables. The code now requires the amount of bracing to meet the greater of that required by the seismic table or the wind table. For the majority of the country, wind loading is the dominant lateral force and controls bracing design. The previous table provided values for both wind and seismic forces but was based predominantly on seismic loads, with wind bracing assumed to be similar to percentages of the seismic values. The resulting bracing amounts to resist design wind speeds of 90 mph to 100 mph were found to be inadequate, and the new wind table reflects increased values for the amount of bracing. The change recognizes that seismic loading is predominantly proportional to the length of the building (i.e., the length of the braced wall line), but wind loading is proportional to the wall line spacing, the height of the walls, and the height of the roof relative to the eaves (i.e., the sail area upon which the wind pressure is exerted). Accordingly, values in the seismic table are a function of braced wall line length, and values in the wind table are a function of braced wall line spacing.

Values in the wind table are based on an assumed 10-foot-high wall for each story and a 10-foot height between the eave and the ridge of the roof. For other than continuous sheathing methods, bracing amounts assume installation of gypsum board on the interior side of the wall. Wind base values are for wind Exposure Category B and a 30-foot mean roof height. Adjustment factors are provided for Exposure Category C and a number of other variables such as wall height.

For buildings in SDCs A and B, and detached one- and two-family dwellings in SDC C, wind load controls and the bracing amounts are found in Table R602.10.1.2(1), Bracing Requirements Based on Wind.

The total amount of bracing required is determined from the applicable wind or seismic table and all applicable adjustment factors. To make this section easier to understand and use, the amount of bracing is now stated as length in feet rather than percentage of the wall.

Section R602.10.1.2 clarifies that for other than angled walls, only walls parallel to the braced wall line count in satisfying the amount of bracing requirements. While walls perpendicular to the braced wall line do contribute to bracing in the perpendicular direction, they do not actively resist the lateral loads in the direction of the wall line being considered and do not therefore contribute to the bracing requirement of that braced wall line.

R602.10.1.2 continues

R602.10.1.2 continued

Where exterior braced wall panels are subjected to wind uplift, connections must be provided unless the weight of the wall above off-sets the wind uplift forces. When the net uplift at the bottom of the wall exceeds 100 plf, connections such as straps must be provided from story-to-story to provide a complete load path from the roof to the foundation. In lieu of the uplift connections required by the code, the code permits the bracing system to be designed (engineered) to resist combined uplift and shear.

Panel type bracing systems that are restrained from overturning have approximately 25% more shear capacity than unrestrained panels of the same type. Therefore, a footnote was added to the wind bracing table that permits the required bracing length for the panel type methods (other than let-in bracing) to be reduced by a factor of 0.80 in one-story or the top story of two- or three-story buildings (1/1.25 = 0.80).

CHANGE TYPE: Addition

CHANGE SUMMARY: This new section allows angled wall segments to contribute to the amount of wall bracing in a braced wall line.

2009 CODE: **R602.10.1.3 Angled Corners.** At corners, braced wall lines shall be permitted to angle out of plane up to 45 degrees with a maximum diagonal length of 8 feet (2438 mm). When determining the length of bracing required, the length of each braced wall line shall be determined as shown in Figure R602.10.1.3. The placement of bracing for the braced wall lines shall begin at the point where the braced wall line, which contains the angled wall adjoins the adjacent braced wall line (Point A as shown in Figure R602.10.1.3.). Where an angled corner is constructed at an angle equal to 45 degrees and the diagonal length is no more than 8 feet (2438 mm), the angled wall may be considered as part of either of the adjoining braced wall lines, but not both. Where the diagonal length is greater than 8 feet (2438 mm), it shall be considered its own braced wall line and be braced in accordance with Section R602.10.1 and methods in Sections R602.10.2 and R602.10.4.

CHANGE SIGNIFICANCE: Angled walls that form other than 90-degree corners are not uncommon in the design of dwellings and additions to dwellings. There were previously no provisions in the code to count angled walls as a continuation of a braced wall line in contributing to the amount of required bracing. The angled portion is capable of resisting a component of the lateral load being transferred along the braced wall line. As the angle from the direction of the braced wall line increases, the capacity of the angled segment to resist the loads decreases. This change permits angled walls up to 8 feet long and no more than 45 degrees out of plane of the braced wall line to be included in the amount of required bracing.

R602.10.1.3
Angled Corners of Braced Wall Lines

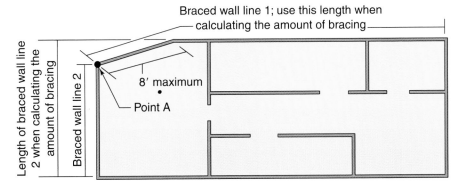

Angled corners of braced wall lines

R602.10.1.4

Braced Wall Panel Location

CHANGE TYPE: Modification

CHANGE SUMMARY: The location requirements for braced wall panels are now grouped together in a single section. The distance from the end of the braced wall line to the first braced wall panel at each end is now limited to a combined total of 12.5 feet. New figures clarify the application of these provisions.

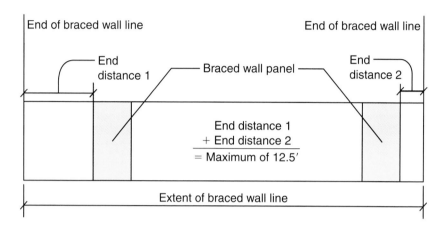

Braced wall panel end distance requirements (SDC A, B, and C)

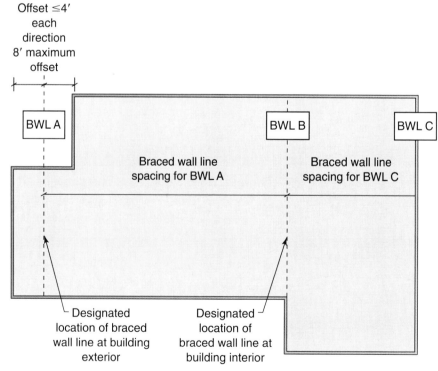

NOTE: Braced wall spacing for BWL B is the greater of the distance from BWL A to BWL B or from BWL B to BWL C.

Braced wall line spacing

2009 CODE: R602.10.1.4 Braced Wall Panel Location.   Braced wall panels shall be located in accordance with Figure R602.10.1.4(1). Braced wall panels shall be located not more than 25 feet on center and shall be permitted to begin no more than 12.5 feet (3810 mm) from the end of a braced wall line in accordance with Section R602.10.1 and Figure R602.10.1.4(2). The total combined distance from each end of a braced wall line to the outermost braced wall panel or panels in the line shall not exceed 12.5 feet. Braced wall panels may be offset out-of-plane up to 4 feet (1219 mm) from the designated braced wall line, provided that the total out-to-out offset of braced wall panels in a braced wall line is not more than 8 feet (2438 mm) in accordance with Figures R602.10.1.3(3) and R602.10.1.3(4). All braced wall panels within a braced wall line shall be permitted to be offset from the designated braced wall line.

CHANGE SIGNIFICANCE: This modification places all panel location requirements together in one section and adds several figures. The spacing of panels is extracted from the 2006 IRC Wall Bracing Table R602.10.1 in the Amount of Bracing column. There is a significant change in the permitted distance from the end of the braced wall line to the first braced wall panel. Historically, braced wall panels have been required at each end of a braced wall line or allowed to be inset a limited distance from the end. The 2006 IRC permitted a maximum inset distance of 12.5 feet from both ends of a braced wall line, provided the amount of bracing satisfied the percentage requirements. The 2009 IRC limits the combined total inset distance to 12.5 feet while still allowing flexibility to inset a panel a distance of up to 12.5 feet from one end. The change is a result of concerns that one 4-foot braced wall panel installed in the center of a 25-foot-long braced wall line (10.5 feet from each end) would not provide adequate bracing even if it satisfied the minimum bracing length requirements. The braced wall panel location requirements for SDCs D_0, D_1, and D_2 have been relocated from Section R602.10.11.2 of the 2006 IRC.

New text clarifies that all of the braced wall panels are allowed to be offset 4 feet from the line that establishes the braced wall line, and the total out-to-out offset of braced wall panels is not more than 8 feet.

R602.10.1.5

Braced Wall Line Spacing for Seismic Design Categories D_0, D_1, and D_2

CHANGE TYPE: Modification

CHANGE SUMMARY: This change expands the exception to permit braced wall line spacing of 35 feet for buildings in Seismic Design Categories D_0, D_1, and D_2 subject to adjustment factors to provide an amount of bracing adjusted to be equivalent to the 25-foot spacing requirements.

2009 CODE: ~~R602.10.11.1~~ <u>R602.10.1.5</u> **Braced Wall Line Spacing for Seismic Design Categories D_0, D_1, and D_2.** Spacing between braced wall lines in each story shall not exceed 25 feet (7620 mm) on center in both the longitudinal and transverse directions.

Exception: In one- and two-story buildings, spacing between two adjacent braced wall lines shall not exceed 35 feet (10 668 mm) on center in order to accommodate one single room not exceeding 900 square feet (84 m^2) in each dwelling unit. Spacing between all other braced wall lines shall not exceed 25 feet (7 620 mm). <u>A spacing of 35 feet (10 668 mm) or less</u>

Example:
Building location:
SDC D_1
Wind 90 MPH
Seismic controls
Minimum bracing length is
determined from Table R602.10.1(2)
based on length of braced wall line

BWL = Braced wall line
BWP = Braced wall panel

Minimum bracing for
25 ft. BWL spacing

Minimum bracing adjusted for
35 ft. BWL spacing per Table R602.10.1.5

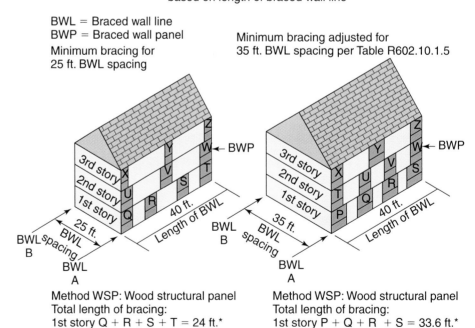

Method WSP: Wood structural panel
Total length of bracing:
1st story Q + R + S + T = 24 ft.*
2nd story U + V + W = 18 ft.*
3rd story X + Y + Z = 8 ft.*

Method WSP: Wood structural panel
Total length of bracing:
1st story P + Q + R + S = 33.6 ft.*
2nd story T + U + V + W = 25.2 ft.*
3rd story X + Y + Z = 11.2 ft.*

*Additional adjustment factors may apply

Braced wall line spacing for Seismic Design Categories D_0, D_1, and D_2

shall be permitted between braced wall lines where the length of wall bracing required by Table R602.10.1.2(2) is multiplied by the appropriate adjustment factor from Table R602.10.1.5, the length-to-width ratio for the floor/roof diaphragm does not exceed 3:1, and the top plate lap splice face nailing is twelve 16d nails on each side of the splice.

CHANGE SIGNIFICANCE: This change expands the exception to the 25-foot braced wall line spacing in Seismic Design Categories D_0, D_1, and D_2 to provide more flexibility in dwelling and townhouse design. The 2006 IRC permitted an increase to 35-foot braced wall line spacing for only one large room not exceeding 900 square feet. The new exception allows spacing up to 35 feet throughout the building by increasing the amount of braced wall panels in the braced wall line. Adjustment factors ensure that the total amount of wall bracing is equivalent to that provided when the braced wall line spacing is limited to 25 feet. This exception also places limits on the length-to-width ratio for the roof and floor diaphragms to ensure that lateral loads are adequately transferred to the braced wall lines and increases the fastening for top plate splices to account for the increased span of the diaphragm. The change does not reduce the seismic resistance but allows the same building plans to be used in Seismic Design Categories D_0, D_1, and D_2 as are used in Seismic Design Category C when the appropriate bracing adjustments are applied.

TABLE R602.10.1.5   **Adjustments of Bracing Length for Braced Wall Lines Greater Than 25 Feet[a,b]**

Braced Wall Line Spacing (feet)	Multiply Bracing Length in Table R602.10.1(2) by:
25	1.0
30	1.2
35	1.4

For SI: 1 foot = 304.8 mm.
a. Linear interpolation is permissible.
b. When a braced wall line has a parallel braced wall line on both sides, the larger adjustment factor shall be used.

R602.10.2

Intermittent Braced Wall Panel Construction Methods

CHANGE TYPE: Modification

CHANGE SUMMARY: The bracing methods of the 2006 IRC listed as types 1 through 8 and the two alternate braced wall panel methods have been grouped into one table and given a two- or three-letter abbreviation to make the section more user friendly. These methods are now considered *intermittent* braced wall panel construction methods to clearly separate them from the *continuous sheathing* methods of bracing. In most cases, ½-inch gypsum wallboard is now required on the side of the wall opposite the bracing material.

2009 CODE: R602.10.2 Intermittent Braced Wall Panel Construction Methods. The construction of intermittent braced wall panels shall be in accordance with one of the methods listed in Table R602.10.2.

R602.10.2.1 Intermittent Braced Wall Panel Interior Finish Material. Intermittent braced wall panels shall have gypsum wall board installed on the side of the wall opposite the bracing material.

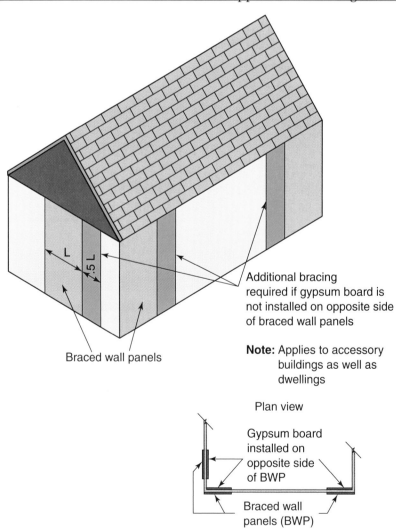

L

.5 L

Additional bracing required if gypsum board is not installed on opposite side of braced wall panels

Braced wall panels

Note: Applies to accessory buildings as well as dwellings

Plan view

Gypsum board installed on opposite side of BWP

Braced wall panels (BWP)

Intermittent braced wall panels

TABLE R602.10.2 **Intermittent Bracing Methods**

Method	Material	Minimum Thickness	Figure	Connection Criteria
LIB	Let-in-bracing	1 × 4 wood or approved metal straps at 45° to 60° angles for 16" stud spacing only		Wood: 2-8d nails per stud Metal: per manufacturer
DWB	Diagonal wood boards at 24" spacing	$5/8$"		2-8d (2½" × 0.113") nails or 2 staples, 1¾" per stud
WSP	Wood structural panel (see Section R604)	$3/8$"		6d common (2" × 0.113") nails at 6" spacing (panel edges) and at 12" spacing (intermediate supports) or 16 ga. × 1¾ staples: at 3" spacing (panel edges) and 6" spacing (intermediate supports)
SFB	Structural fiberboard sheathing	½" or $25/32$" for 16" stud spacing only		1½" galvanized roofing nails or 8d common (2½" × 0.131) nails at 3" spacing (panel edges) at 6" spacing (intermediate supports)
GB	Gypsum board	½"		Nails or screws at 7" spacing at panel edges including top and bottom plates; for exterior sheathing nail size, see Table R602.3(1); for interior gypsum board nail size, see Table R702.3.5
PBS	Particleboard sheathing	$3/8$" or ½" for 16" stud spacing only		1½" galvanized roofing nails or 8d common (2½" × 0.131) nails at 3" spacing (panel edges) at 6" spacing (intermediate supports)
PCP	Portland cement plaster	See Section R703.6		1½", 11 gage, $7/16$" head nails at 16" spacing or $7/16$", 16 gage staples at 6" spacing
HPS	Hardboard panel siding	$7/16$"		0.092" dia., 0.225" head nails with length to accommodate 1½" penetration into studs at 4" spacing (panel edges), at 8" spacing (intermediate supports)
ABW	Alternate braced wall	See Section R602.10.3.2		See Section R602.10.3.2
PFH	Intermittent portal frame	See Section R602.10.3.3		See Section R602.10.3.3
PFG	Intermittent portal frame at garage	See Section R602.10.3.4		See Section R602.10.3.4

For SI: 1 inch = 25.4 mm; 1 foot = 305 mm.

R602.10.2 continues

R602.10.2 continued Gypsum wall board shall be not less than ½ inch (12.7 mm) in thickness and be fastened in accordance with Table R702.3.5 for interior gypsum wall board.

Exceptions:

1. Wall panels that are braced in accordance with Methods GB, ABW, PFG and PFH.

2. When an approved interior finish material with an in-plane shear resistance equivalent to gypsum board is installed.

3. For Methods DWB, WSP, SFB, PBS, PCP, and HPS, omitting gypsum wall board is permitted provided the length of bracing in Tables R602.10.1.2(1) and R602.10.1.2(2) is multiplied by a factor of 1.5.

CHANGE SIGNIFICANCE: The code now uses the term "intermittent" to describe bracing methods utilizing isolated braced wall panels and to clearly differentiate these methods from the continuous sheathing methods. The intermittent bracing methods are placed in tabular format with a description, illustrative icon, and connection criteria. Short abbreviations are used in place of numbers for each bracing type. For example, Method 3 using wood structural panels is now designated as method WSP. The new tabular format is intended to make it easier for code users to understand the options available. The two alternate bracing methods are added into the table and also given abbreviations ABW and PFH. A new alternate for intermittent portal frames at garage door openings in Seismic Design Categories A, B and C (Method PFG) completes the list. The reorganization and labeling intend to clarify the prescriptive bracing provisions and the two distinct paths for compliance—intermittent and continuous methods—to promote more consistent application.

Comparison of Braced Wall Panel Construction Method Designations in the 2006 and 2009 Editions of the IRC

Bracing Description	2006 IRC Bracing Method Designation	2009 IRC Bracing Method Designation
Let-in-bracing	1	LIB
Diagonal wood boards	2	DWB
Wood structural panel	3	WSP
Structural fiberboard sheathing	4	SFB
Gypsum board	5	GB
Particleboard sheathing	6	PBS
Portland cement plaster	7	PCP
Hardboard panel siding	8	HPS
Alternate braced wall	Alternate braced wall panel construction (Section R602.10.6.1)	ABW
Intermittent portal frame	Alternate braced wall panel adjacent to a door or window opening (Section R602.10.6.2)	PFH

It is common in dwelling construction to install ½-inch gypsum board as finish material for interior walls. In effect, this gypsum board installed on the side of the wall opposite the bracing materials provides additional stiffness and resistance to lateral loads. Through code change proposals and testimony at code hearings, proponents have asserted that the required amounts of wall bracing are based on the installation of gypsum board on the opposite side of the braced wall panel, and without it the braced wall line has insufficient load-resisting capacity. The addition of gypsum board as a specific requirement is to clarify the intent and ensure installation of this component to provide adequate wall bracing. The exceptions recognize alternative equivalent materials, or in the absence of interior materials as part of the bracing system, require an increase to the amount of bracing. When gypsum board (Method GB) is the bracing method, either single-sided or double-sided application is already recognized in the provisions. The alternative bracing methods, ABW and PFH, are stand-alone provisions that do not require interior gypsum board finish.

The prohibition of adhesive attachment of wall sheathing in Seismic Design Categories C, D_0, D_1, and D_2 is relocated from Section R602.10.11.5 of the 2006 IRC. While this section serves to emphasize the importance of not relying on adhesive connections in higher seismic areas, the code does not include adhesive attachment for any of the bracing methods including gypsum board regardless of the applicable seismic or wind speed classification. The explicit prohibition means that even proprietary adhesive fastening systems would not be permitted to attach braced wall sheathing in Seismic Design Categories C, D_0, D_1, and D_2.

R602.10.3

Minimum Length of Braced Wall Panels

CHANGE TYPE: Modification

CHANGE SUMMARY: For most methods of bracing, the code now recognizes braced wall panels less than 48 inches but not less than 36 inches in length in Seismic Design Categories A, B, and C. Partial credit, expressed in effective length, is given to these panels in contributing to the strength of the braced wall line, provided the height to length aspect ratio does not exceed 2.5 to 1. When gypsum board panels (Method GB) are applied to only one face of a braced wall panel, the required bracing length must be doubled.

2009 CODE: ~~R602.10.4~~ <u>R602.10.3</u> **Minimum Length of Braced Panels.** For Methods ~~2, 3, 4, 6, 7 and 8 above~~ <u>DWB, WSP, SFB, PBS, PCP, and HPS</u>, each braced wall panel shall be at least 48 inches (1219 mm) in length, covering a minimum of three stud spaces where studs are spaced 16 inches (406 mm) on center and covering a minimum of two stud spaces where studs are spaced 24 inches (610 mm) on center. For Method ~~5 above~~ <u>GB</u>, each braced wall panel shall be

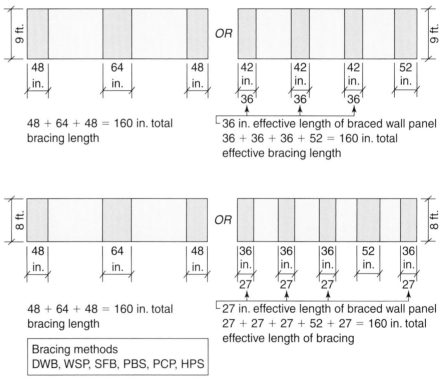

48 + 64 + 48 = 160 in. total
bracing length

36 in. effective length of braced wall panel
36 + 36 + 36 + 52 = 160 in. total
effective bracing length

48 + 64 + 48 = 160 in. total
bracing length

Bracing methods
DWB, WSP, SFB, PBS, PCP, HPS

27 in. effective length of braced wall panel
27 + 27 + 27 + 52 + 27 = 160 in. total
effective length of bracing

**Equivalent effective bracing lengths using partial credit
for braced wall panels 42 in. and 36 in. in length**

at least 96 inches (2438 mm) in length where applied to one face of a braced wall panel and at least 48 inches (1219 mm) where applied to both faces. For Methods DWB, WSP, SFB, PBS, PCP, and HPS, for purposes of computing the length of panel bracing required in Tables R602.10.1.2 (1) and R602.10.1.2 (2), the effective length of the braced wall panel shall be equal to the actual length of the panel. When GB panels are applied to only one face of a braced wall panel, bracing length required in Tables R602.10.1.2 (1) and R602.10.1.2 (2) for Method GB shall be doubled.

Exceptions:

1. Lengths of braced wall panels for continuous ~~wood structural panel~~ sheathing methods shall be in accordance with ~~Section R602.10.5~~ Section R602.10.4.2.

2. Lengths of ~~alternate braced wall~~ Method ABW panels shall be in accordance with Section ~~R602.10.6.1 or Section R602.10.6.2~~ R602.10.3.2.

3. Length of Methods PFH and PFG shall be in accordance with Section R602.10.3.3 and R602.10.3.4 respectively.

4. For Methods DWB, WSP, SFB, PBS, PCP, and HPS in Seismic Design Categories A, B, and C: Panels between 36 inches and 48 inches in length shall be permitted to count toward the required length of bracing in Tables R602.10.1.2(1) and R602.10.1.2(2), and the effective contribution shall comply with Table R602.10.3.

CHANGE SIGNIFICANCE: As part of the extensive revision of the wall bracing provisions of Section R602.10, the amount of bracing is now expressed as the minimum total length of braced wall panels measured in the direction of the braced wall line. Previously the code was inconsistent in terminology, expressing such braced wall panel measurements as width, length, or a percentage of the braced wall line. In most cases, the length of the braced wall panel in the 2009 IRC is equal to the actual length of the panel in the horizontal direction, provided it is not less than 48 inches. When the method of bracing is gypsum board (GB), the new tables R602.10.1.2 (1), Bracing Requirements Based on Wind, and R602.10.1.2 (2), Bracing Requirements Based on Seismic Design Category, assume that gypsum board is applied to

TABLE R602.10.3   Effective Lengths for Braced Wall Panels Less Than 48 Inches in Actual Length (Brace Methods DWB, WSP, SFB, PBS, PCP, and HPS[a])

Actual Length of Braced Wall Panel (inches)	Effective Length of Braced Wall Panel (inches)		
	8-foot Wall Height	9-foot Wall Height	10-foot Wall Height
48	48	48	48
42	36	36	N/A
36	27	N/A	N/A

For SI: 1 inch = 25.4 mm, 1 foot = 304.8 mm.
Interpolation shall be permitted.

R602.10.3 continues

R602.10.3 continued

both sides of the wall. The added text in Section R602.10.3 mirrors the footnotes of the referenced tables in advising that the amount of gypsum board required by the applicable table must be doubled when gypsum board is applied to only one side. This is consistent with the previous statement in the section that states the minimum length of a gypsum braced wall panel is 48 inches when applied to both sides and 96 inches when applied to only one side.

For intermittent braced wall panels using methods other than let-in-bracing and gypsum board, the code now recognizes that panels less than 48 inches in length contribute to the bracing of buildings. For braced wall panels not less than 36 inches in length in Seismic Design Categories A, B, and C, the new partial credit allowance maintains the bracing strength requirements while providing some flexibility for architectural design. The effective lengths of braced wall panels shown in Table R602.10.3 account for reduction in bracing strength relative to a standard 48-inch braced wall panel in proportion to increase in overturning forces on conventional connections at the base of the wall. Partial credit is permitted only where the aspect ratio of panel height to panel length does not exceed 2.5 to 1. This aspect ratio limitation corresponds to a standard 48-inch braced wall panel with a 10-foot wall height. Therefore, there is no partial credit for panels less than 48 inches in length on 10-foot-high walls or panels less than 42 inches in length on 9-foot-high walls.

CHANGE TYPE: Modification

CHANGE SUMMARY: A new figure replaces much of the text in this section to more clearly illustrate the construction details for alternate braced wall panels, now described as bracing method ABW.

2009 CODE: ~~R602.10.6 Alternate Braced Wall Panel Construction Methods.~~ ~~Alternate braced wall panels shall be constructed in accordance with Sections R602.10.6.1 and R602.10.6.2.~~

~~R602.10.6.1~~ R602.10.3.2 Method ABW: Alternate Braced Wall Panels. ~~Alternate braced wall lines~~ Method ABW braced wall panels constructed in accordance with one of the following provisions shall be permitted to replace each 4 feet (1219 mm) of braced wall panel as required by Section ~~R602.10.4~~ R602.10.3. The maximum height and minimum ~~width~~ length and hold-down force of each panel shall be in accordance with Table ~~R602.10.6~~ R602.10.3.2:

1. In one-story buildings, each panel shall be ~~sheathed on one face with 3/8-inch-minimum-thickness (10 mm) wood structural panel sheathing nailed with 8d common or galvanized box nails in accordance with Table R602.3(1) and blocked at all wood structural panel sheathing edges.~~ installed in accor-

R602.10.3.2
Method ABW: Alternate Braced Wall Panels

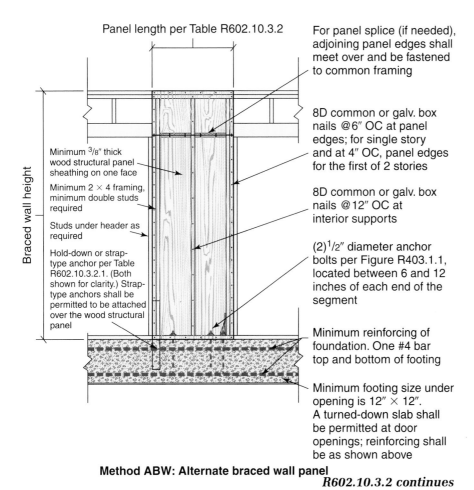

Panel length per Table R602.10.3.2

For panel splice (if needed), adjoining panel edges shall meet over and be fastened to common framing

8D common or galv. box nails @ 6″ OC at panel edges; for single story and at 4″ OC, panel edges for the first of 2 stories

8D common or galv. box nails @ 12″ OC at interior supports

(2)1/2″ diameter anchor bolts per Figure R403.1.1, located between 6 and 12 inches of each end of the segment

Minimum reinforcing of foundation. One #4 bar top and bottom of footing

Minimum footing size under opening is 12″ × 12″. A turned-down slab shall be permitted at door openings; reinforcing shall be as shown above

Braced wall height

Minimum 3/8″ thick wood structural panel sheathing on one face

Minimum 2 × 4 framing, minimum double studs required

Studs under header as required

Hold-down or strap-type anchor per Table R602.10.3.2.1. (Both shown for clarity.) Strap-type anchors shall be permitted to be attached over the wood structural panel

Method ABW: Alternate braced wall panel

R602.10.3.2 continues

R602.10.3.2 continued

dance with Figure R602.10.3.2. ~~Two anchor bolts installed in accordance with Figure R403.1(1) shall be provided in each panel. Anchor bolts shall be placed at panel quarter points. Each panel end stud shall have a tie-down device fastened to the foundation, capable of providing an uplift capacity in accordance with Table R602.10.6.~~ The ~~tie~~ <u>hold</u>-down device shall be installed in accordance with the manufacturer's recommendations. The panels shall be supported directly on a foundation or on floor framing supported directly on a foundation, which is continuous across the entire length of the braced wall line. ~~This foundation shall be reinforced with not less than one No. 4 bar top and bottom. When the continuous foundation is required to have a depth greater than 12 inches (305 mm), a minimum 12-inch-by-12-inch (305 mm by 305 mm) continuous footing or turned down slab edge is permitted at door openings in the braced wall line. This continuous footing or turned down slab edge shall be reinforced with not less than one No. 4 bar top and bottom. This reinforcement shall be lapped 15 inches (381 mm) with the reinforcement required in the continuous foundation located directly under the braced wall line.~~

2. In the first story of two-story buildings, each braced wall panel shall be in accordance with Item 1 above, except that the wood structural panel ~~sheathing shall be installed on both faces,~~ sheathing edge nailing spacing shall not exceed ~~4~~ <u>four</u> inches (102 mm) on center~~, at least three anchor bolts shall be placed at one-fifth points~~.

~~TABLE R602.10.6~~ <u>TABLE R602.10.3.2</u>

Minimum ~~Widths~~ <u>Length Requirements</u> and ~~Tie~~ <u>Hold</u>-Down Forces ~~of~~ ~~Alternate~~ <u>For Method ABW</u> Braced Wall Panels

Seismic Design Category and Wind Speed		Height of Braced Wall Panel				
		8 ft.	9 ft.	10 ft.	11 ft.	12 ft.
SDCs A, B, and C Wind speed <110 mph	Minimum sheathed ~~Width~~ <u>length</u>	2'-4"	2'-8"	2'-10"	3'-2"	3'-6"
	R602.10.3.2, Item 1 ~~Tie~~ <u>Hold</u>-down Force (lbs)	1800	1800	1800	2000	2200
	R602.10.3.2, Item 2 ~~Tie~~ <u>Hold</u>-down Force (lbs)	3000	3000	3000	3300	3600
SDCs D_0, D_1, and D_2 Wind speed <110 mph	Minimum sheathed length	2'-8"	2'-8"	2'-10"	NP[a]	NP[a]
	R602.10.3.2, Item 1 ~~Tie~~ <u>Hold</u>-down Force (lbs)	1800	1800	1800	NP[a]	NP[a]
	R602.10.3.2, Item 2 ~~Tie~~ <u>Hold</u>-down Force (lbs)	3000	3000	3000	NP[a]	NP[a]

For SI: 1 inch = 25.4 mm; 1 foot = 305 mm; 1 pound = 4.448 N.
a. NP = Not permitted. Maximum height of 10 feet.

CHANGE SIGNIFICANCE: Alternate braced wall panel construction, now known as method ABW, is one of the more complicated provisions in the wall bracing section. The construction details for minimum materials, concrete reinforcement, hold-downs, anchoring, fastening, and splicing are more clearly illustrated in drawing form rather than detailed code language. Much of the text of this section has been deleted in favor of the new figure without making technical changes to the method of construction. The intent is to promote better understanding of the provisions and consistency in their application for both the builder and the inspector.

Other editorial changes to this section reflect the preferred terminology in an effort to provide accuracy and consistency. The horizontal dimension for braced wall panels is now consistently expressed as length instead of width. Hold-down replaces tie-down in describing the anchoring device and the minimum load-resisting capacity of the device.

R602.10.3.3

Method PFH: Portal Frame with Hold-Downs

CHANGE TYPE: Modification

CHANGE SUMMARY: The alternate bracing method for a braced wall panel adjacent to a door or window opening, typically used at large overhead garage door openings, is now known as *portal frame with hold-downs* (Method PFH). The text describing the materials and connection details has been deleted in favor of Figure R602.10.3.2 for illustrating this method of bracing construction.

2009 CODE: ~~R602.10.6.2 Alternate Braced Wall Panel Adjacent to a Door or Window Opening.~~ <u>R602.10.3.3 Method PFH: Portal Frame with Hold-Downs.</u> ~~Alternate~~ <u>Method PFH</u> braced wall panels constructed in accordance with one of the following provisions are also permitted to replace each 4 feet (1219 mm) of braced wall panel as required by Section ~~R602.10.4~~ <u>R602.10.3</u> for use adjacent to a window or door opening with a full-length header:

1. ~~In one-story buildings, each panel shall have a length of not less than 16 inches (406 mm) and a height of not more than 10 feet (3048 mm).~~ Each panel ~~sheathed on one face with a single layer of 3/8-inch-minimum-thickness (10 mm) wood structural panel sheathing nailed with 8d common or galvanized box nails in accordance with Figure R602.10.6.2.~~ <u>shall be fabricated in accordance with Figure R602.10.3.3.</u> The wood structural panel sheathing shall extend up over the solid sawn or glued-laminated header and shall be nailed in accordance with Figure ~~R602.10.6.2~~ <u>R602.10.3.3.</u> ~~Use of a built-up header consisting of at least two 2 x 12s and fastened in accordance with Table~~

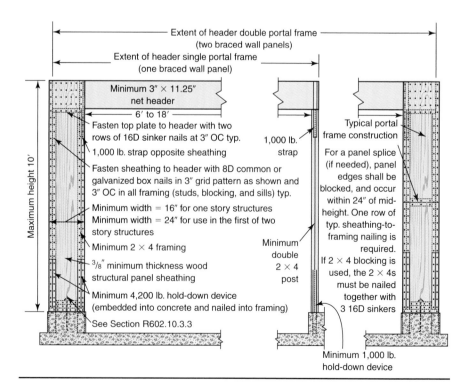

Method PFH: Portal frame with hold-downs

~~R602.3(1) shall be permitted.~~ A spacer, if used <u>with built-up header</u>, shall be placed on the side of the built-up beam opposite the wood structural panel sheathing. The header shall extend between the inside faces of the first full-length outer studs of each panel. ~~The header shall extend between the inside faces of the first full-length outer studs of each panel. The clear span of the header between the inner studs of each panel shall be not less than 6 feet (1829 mm) and not more than 18 feet (5486 mm) in length. A strap with an uplift capacity of not less than 1000 pounds (4448 N) shall fasten the header to the side of the inner studs opposite the sheathing.~~ One anchor bolt not less than ⅝-inch diameter (16 mm) and installed in accordance with Section R403.1.6 shall be provided in the center of each sill plate. ~~The studs at each end of the panel shall have a tie-down device fastened to the foundation with an uplift capacity of not less than 4,200 pounds (18 683 N).~~

~~Where a panel is located on one side of the opening, the header shall extend between the inside face of the first full-length stud of the panel and the bearing studs at the other end of the opening. A strap with an uplift capacity of not less than 1000 pounds (4448 N) shall fasten the header to the bearing studs. The bearing studs shall also have a tie-down device fastened to the foundation with an uplift capacity of not less than 1000 pounds (4448 N).~~

The ~~tie~~ <u>hold</u>-down devices shall be an embedded-strap type, installed in accordance with the manufacturer's recommendations. The panels shall be supported directly on a foundation, which is continuous across the entire length of the braced wall line. The foundation shall be reinforced ~~with not less than one No. 4 bar top and bottom~~ <u>as shown in Figure R602.10.3.2.</u> ~~Where the continuous foundation is required to have a depth greater than 12 inches (305 mm), a minimum 12-inch-by-12-inch (305 mm by 305 mm) continuous footing or turned down slab edge is permitted at door openings in the braced wall line. This continuous footing or turned down slab edge shall be reinforced with not less than one No. 4 bar top and bottom.~~ This reinforcement shall be lapped not less than 15 inches (381 mm), with the reinforcement required in the continuous foundation located directly under the braced wall line.

2. In the first story of two-story buildings, each wall panel shall be braced in accordance with Item 1 above, except that each panel shall have a length of not less than 24 inches (610 mm).

CHANGE SIGNIFICANCE: The alternate for braced wall panels adjacent to a door or window opening first appeared in the 2006 IRC. Developed through testing by APA Laboratory, this method is used primarily to reduce the length of the braced panel at the sides of large overhead garage door openings. Builders and inspectors have relied on the detailed information provided by Figure R602.10.6.2 in understanding this bracing method. The braced wall segments in the figure are labeled as "portal frames," a term that more accurately describes the configuration, prompting a change to designate this method of brac-

R602.10.3.3 continues

R602.10.3.3 continued

ing as *portal frame with hold-downs* (Method PFH). As with alternate braced wall panels (ABW), bracing method PFH provides equivalent strength to a standard 48-inch braced wall panel through very specific reinforcing and connection details. The lengthy text describing those details was viewed as cumbersome and confusing and has been deleted in favor of the line drawing illustration. The construction details for minimum materials, concrete reinforcement, hold-downs, anchoring, fastening, and splicing are more clearly illustrated in the figure. The illustration includes a minor revision to clarify that two anchor bolts are required at the portal frame panel.

CHANGE TYPE: Modification

CHANGE SUMMARY: The continuous sheathing method of bracing has undergone extensive revision and expansion to provide more flexibility in the design and construction of dwellings. The code now recognizes the practice of mixing intermittent bracing methods with the continuous sheathing method. In SDC A, B and C where the basic wind speed is less than or equal to 100 mph, the code permits mixing of methods in the same story and from story to story. When using the continuous sheathing method in Seismic Design Categories D_0, D_1, and D_2 or where the wind speed exceeds 100 mph, mixing is not permitted on the same story.

The total length of required bracing in a braced wall line appears in the applicable column of Table R602.10.1.2(1), when wind controls, and Table R602.10.1.2(2), when seismic controls, and is no longer related to the adjacent clear opening height.

R602.10.4
Continuous Sheathing

2009 CODE: ~~R602.10.5~~ R602.10.4 **Continuous Sheathing.** Braced wall lines with continuous sheathing shall be constructed in accordance with this section. All braced wall lines along exterior walls on the same story shall be continuously sheathed.

> **Exception:** Within Seismic Design Categories A, B, and C or in regions where the basic wind speed is less than or equal to 100 mph, other bracing methods prescribed by this code shall be permitted on other braced wall lines on the same story level or on any braced wall line on different story levels of the building.

R602.10.4.1 Continuous Sheathing Braced Wall Panels. Continuous sheathing methods require structural panel sheathing to be used

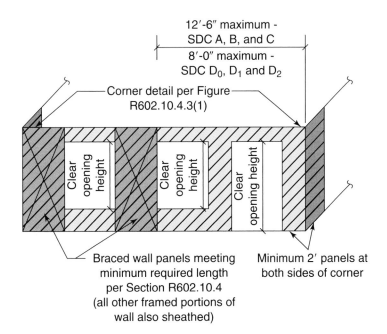

Continuously sheathed braced wall line — first braced wall panel away from end of wall line without tie down

R602.10.4 continues

R602.10.4 continued on all sheathable surfaces on one side of a braced wall line, including areas above and below openings and gable end walls. Braced wall panels shall be constructed in accordance with one of the methods listed in Table R602.10.4.1. Different bracing methods, other than those listed in Table R602.10.4.1, shall not be permitted along a braced wall line with continuous sheathing.

Because numerous code changes deleted, added, and modified substantial portions of Section R602.10, the entire code change text is too extensive to be included here. Refer to the *2009 IRC Code Changes Resource Collection* for the complete text and history of code changes related to Section R602.10.

CHANGE SIGNIFICANCE: In an effort to clearly differentiate intermittent from continuous bracing methods, the continuous sheathing provisions are no longer tied to wood structural panel bracing method WSP (formerly Method 3). The previous language caused some confusion and did not clearly explain that continuous sheathing was a separate path from isolated wood structural panels for compliance with the bracing provisions. Continuous sheathing requires wood structural panels on all sheathable surfaces on one side of the entire braced wall line. Because this method creates a stronger, stiffer, and more structurally redundant wall, required braced wall panel lengths are reduced accordingly when compared with intermittent braced wall panels.

Previously, Table R602.10.5 computed the panel length required based on the height of the adjacent door or window and the applicable maximum height to length aspect ratio. Table R602.10.5 has been deleted in the 2009 IRC, and the minimum total length of braced wall panels for continuous sheathing appears in the applicable column of either Table R602.10.1.2(1), when wind controls, or Table R602.10.1.2(2), when seismic controls. The tabular value is no longer based on adjacent opening heights expressed as a percentage of wall heights. This simplified approach recognizes that some of the percentage values in the 2006 IRC were overly conservative, and a single value for total braced wall panel lengths for continuous sheathing is justified.

TABLE R602.10.4.1   **Continuous Sheathing Methods**

Method	Material	Minimum Thickness	Figure	Connection Criteria
CS-WSP	Wood structural panel	3/8"		6d common (2" × 0.113") nails at 6" spacing (panel edges) and at 12" spacing (intermediate supports) or 16 ga. × 1¾ staples: at 3" spacing (panel edges) and 6" spacing (intermediate supports)
CS-G	Wood structural panel adjacent to garage openings and supporting roof load only[a,b]	3/8"		See Method CS-WSP
CS-PF	Continuous portal frame	See Section R602.10.4.1.1		See Section R602.10.4.1.1

For SI: 1 inch = 25.4 mm.
a. Applies to one wall of a garage only.
b. Roof covering dead loads shall be 3 psf (0.14 kN/m²) or less.

Consistent with other modifications throughout Section R602.10, amounts of required bracing are expressed as the length of braced wall panels in feet rather than a percentage of the braced wall line. Bracing amounts may be reduced with the increased fastening specified in Section R602.10.4.3.

The expanded Section R602.10.4 establishes three separate and distinct methods for bracing with continuous sheathing and assigns helpful abbreviations in a new Table R602.10.4.1 with a similar format to the table for intermittent sheathing methods. The alternates for wood structural panel adjacent to garage openings (CS-G) and continuous portal frame (CS-PF) were developed from the footnotes that appeared in the 2006 IRC Table R602.10.5.

The new provisions for continuous sheathing provide more flexibility for the design of dwellings. Previously, these provisions of the code required wood structural panel sheathing on all sheathable surfaces on one side of all braced wall lines, interior and exterior, of the building. By comparison, Section R602.10.4 of the 2009 IRC requires continuous wood structural panel sheathing on all sheathable surfaces on one side of braced wall lines of exterior walls. This change permits other bracing methods to be used at other braced wall lines at any story. In addition, in the lower Seismic Design Categories (SDCs A, B, and C) and regions with a design wind speed not greater than 100 mph, mixing of bracing methods is permitted from one braced wall line to the next and from one story to the next. It should be noted that continuous wall sheathing cannot be mixed with other bracing methods within the same braced wall line under any conditions.

The code clarifies the requirement for a minimum 24-inch wood structural panel on both sides of the corner at each end of the continuously sheathed braced wall line. A hold-down device with a capacity of 800 pounds installed on the corner stud of the end panel of the braced wall line that provides overturning restraint is permitted to substitute for the 24-inch corner return segment that is perpendicular to the braced wall line.

The continuous portal frame, Method CS-PF, is the result of extensive cyclic testing which demonstrated load resistance performance exceeding other permitted methods of bracing. Previously, footnote *c* of Table R602.10.5 permitted this narrow wall segment with a 6 to 1 aspect ratio and referenced the figure for portal frame construction. The requirements have been moved into the text and reference a new figure clarifying this application. This change does not limit the use of the 6:1 portal frame to a concrete foundation or to Seismic Design Categories A, B, and C. Method CS-PF can now be used over raised floor wood framing, on any story and in all Seismic Design Categories A–D. Three separate 16-inch-wide segments are required to equal a single 48-inch-wide segment or two 24-inch-wide segments. Testing has shown that in many cases the three separate panels perform better than a single 48-inch panel. When using the continuous portal frame method, the total amount of bracing in the braced wall line must still meet the applicable tabular values for continuous wall sheathing. Additional strap requirements for wind Exposure Categories C and D are provided in Table R602.10.4.1.1. These expanded provisions provide additional flexibility in the design and construction of dwellings.

R602.10.6 and R602.10.7

Braced Wall Panel Connections and Support

CHANGE TYPE: Modification

CHANGE SUMMARY: Requirements for braced wall panel connection to wood framing above and below have been revised. The code now recognizes masonry stem wall construction for supporting braced wall panels and prescribes reinforcing when these walls are 48 inches or less in length. It should be noted that alternate braced wall panels and portal frame panels with hold-downs are specifically prohibited from being supported by masonry stem walls.

2009 CODE: **R602.10.6 Braced Wall Panel Connections.** Braced wall panels shall be connected to floor framing or foundations as follows:

1. Where joists are perpendicular to a braced wall panel above or below, a rim joist, band joist, or blocking shall be provided

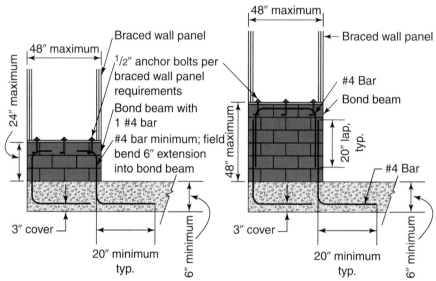

Short stem wall reinforcement   Tall stem wall reinforcement

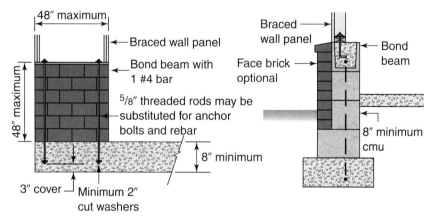

Optional stem wall reinforcement   Typical stem wall section

Note: Grout bond beams and all cells that contain rebar, threaded rods and anchor bolts.

Masonry stem walls supporting braced wall panels

along the entire length of the braced wall panel in accordance with Figure R602.10.6(1). Fastening of top and bottom wall plates to framing, rim joist, band joist, and/or blocking shall be in accordance with Table R602.3(1).

2. Where joists are parallel to a braced wall panel above or below, a rim joist, end joist, or other parallel framing member shall be provided directly above and below the braced wall panel in accordance with Figure R602.10.6(2). Where a parallel framing member cannot be located directly above and below the panel, full-depth blocking at 16-inch (406 mm) spacing shall be provided between the parallel framing members to each side of the braced wall panel in accordance with Figure R602.10.6(2). Fastening of blocking and wall plates shall be in accordance with Table R602.3(1) and Figure R602.10.6(2).

3. Connections of braced wall panels to concrete or masonry shall be in accordance with Section R403.1.6.

R602.10.7 Braced Wall Panel Support.   Braced wall panel support shall be provided as follows:

1. Cantilevered floor joist, supporting braced wall lines, shall comply with Section R502.3.3. Solid blocking shall be provided at the nearest bearing wall location. In Seismic Design Categories A, B, and C, where the cantilever is not more than 24 inches (607 mm), a full height rim joist instead of solid blocking shall be provided.

2. Elevated post or pier foundations supporting braced wall panels shall be designed in accordance with accepted engineering practice.

3. Masonry stem walls with a length of 48 inches (1220 mm) or less supporting braced wall panels shall be reinforced in accordance with Figure R602.10.7. Masonry stem walls with a length greater than 48 inches (1220 mm) supporting braced wall panels shall be constructed in accordance with Section R403.1. Braced wall panels constructed in accordance with Sections R602.10.3.2 and R602.3.3 shall not be attached to masonry stem walls.

CHANGE SIGNIFICANCE: The revised sections more clearly differentiate between connections and supports, clarify the existing connection requirements, and provide details for braced wall panel connection to wood framing. New provisions address connections of rafters and roof trusses to the top of the wall at braced wall panel locations. The new connection details apply to buildings located in SDC D_0, D_1, or D_2 or areas with wind speeds of 100 mph or greater, or when the roof member heel height exceeds 9¼ inches. The change also clarifies that these bracing connection requirements apply to the individual braced wall panel segments, not the entire braced wall line. New figures illustrate the connection options to ensure proper installation without compromising the lateral load resisting capacity.

New provisions in Section R602.10.6 provide direction for support of braced wall panels on masonry stem walls. Masonry stem walls

R602.10.6 and R602.10.7 continues

R602.10.6 and R602.10.7 continued

were not addressed in relation to bracing in previous editions of the IRC. In particular, the absence of language addressing portal frame panels supported by masonry stem walls, as sometimes occur at garage doors and slab-on-grade conditions, has resulted in inconsistent application of the code. The new stem wall reinforcement requirements for braced wall panels less than 48 inches were developed jointly by members of the ICC Ad Hoc Committee on Wall Bracing and the National Concrete Masonry Association. Their findings concluded that the standard provisions for masonry wall construction of Chapter 4 were inadequate to resist the loads associated with a narrow braced wall panel less than 48 inches in length. The masonry reinforcing requirements were determined by computing sample loading conditions and analyzing the effects on the masonry. To provide flexibility in design and construction, optional conditions are provided in the associated figure. Alternate braced wall panels (Method ABW) and portal frame with hold-downs (Method PFH) are not permitted to be supported on masonry stem walls. Analysis concluded that it was not feasible to construct these alternate bracing methods atop masonry and achieve the required load capacity associated with cast-in-place concrete stem walls with hold-downs.

New text also recognizes floor joist cantilever conditions to support braced wall panels consistent with the requirements of Section R502.3.3.

CHANGE TYPE: Modification

CHANGE SUMMARY: The exception permitting horizontal joints without blocking in lower Seismic Design Categories has been deleted. The code now permits horizontal joints without blocking for panel sheathing except hardboard panel siding, provided that the minimum required amount of bracing is doubled. Gypsum board braced wall panels no longer require blocking behind horizontal joints.

2009 CODE: **R602.10.8 Panel Joints.** All vertical joints of panel sheathing shall occur over and be fastened to common studs. Horizontal joints in braced wall panels shall occur over and be fastened to common blocking of a minimum 1½-inch (38 mm) thickness.

> **Exception:** Blocking is not required behind horizontal joints in Seismic Design Categories A and B and detached dwellings in Seismic Design Category C when constructed in accordance with Section R602.10.3, braced-wall panel construction method 3 and Table R602.10.1, method 3, or where permitted by the manufacturer's installation requirements for the specific sheathing material.

R602.10.8

Braced Wall Panel Joints

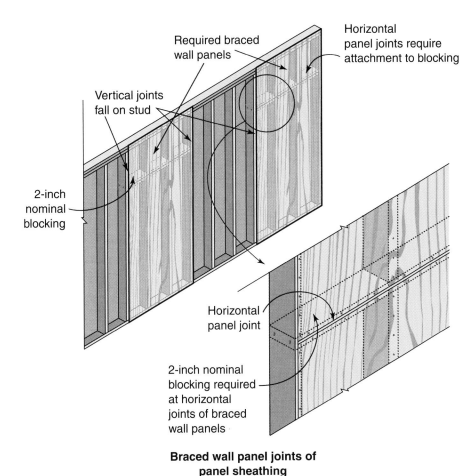

Braced wall panel joints of panel sheathing

R602.10.8 continues

R602.10.8 continued

Exceptions:

1. Blocking at horizontal joints shall not be required in wall segments that are not counted as braced wall panels.

2. Where the bracing length provided is at least twice the minimum length required by Tables R602.10.1.2(1) and R602.10.1.2(2), blocking at horizontal joints shall not be required in braced wall panels constructed using Method WSP, SFB, GB, PBS, or HPS.

3. When Method GB panels are installed horizontally, blocking of horizontal joints is not required.

CHANGE SIGNIFICANCE: Blocking at intermediate joints increases stiffness to keep braced wall panels from buckling out of plane when subjected to in-plane loads. Testing has shown a 50 percent reduction in the bracing strength of wood structural panels when the blocking is omitted behind horizontal joints. As a result, blocking is now required for the horizontal joints of braced wall panel sheathing in all Seismic Design Categories. This change also clarifies that blocking is required only for the prescribed braced wall panels, not the entire braced wall line. Installing twice the amount of otherwise required bracing provides equivalent strength, and the blocking may be omitted. This exemption applies to all approved sheathing panels.

This change clarifies that gypsum board braced wall panels (Method GB) applied horizontally do not require attachment to horizontal blocking at the joints. Proponents of this change stated that taped joints provide adequate shear transfer between panels when the gypsum board is installed horizontally and tape is installed on the joints.

CHANGE TYPE: Modification

R602.10.9
Cripple Wall Bracing

CHANGE SUMMARY: This section has been relocated and the terminology updated to be consistent with other changes to Section R602.10. Required bracing is now measured as length in feet rather than a percentage of the braced wall line and is determined from the wind or seismic table, whichever is the greater value.

2009 CODE: ~~**R602.10.2**~~ **R602.10.9 Cripple Wall Bracing.** ~~**R602.10.2.1 Seismic Design Categories Other than D.**~~ In Seismic Design Categories other than D_2, cripple walls shall be braced with ~~an amount~~ a length and type of bracing as required for the wall above in accordance with ~~Table R602.10.1~~ Tables R602.10.1.2(1) and R602.10.1.2(2) with the following modifications for cripple wall bracing:

1. ~~The percent bracing amount as determined from Table R602.10.1 shall be increased by 15 percent~~ The length of bracing as determined from Tables R602.10.1.2(1) and R602.10.1.2(2) shall be multiplied by a factor of 1.15, and

2. The wall panel spacing shall be decreased to 18 feet (5486 mm) instead of 25 feet (7620 mm).

CHANGE SIGNIFICANCE: As part of the general reorganization and updating of the wall bracing provisions, cripple wall bracing requirements have been grouped into one location in Section R602.10.9 and updated for consistency with the other bracing provisions. Amounts

R602.10.9 continues

Maximum 25 ft.

Wall framing

Required braced wall panels for a braced wall line equal in length to the cripple wall
A + B = 1.15 (Y + Z)
Example :
If Y + Z = 96 in.
then
A + B = 1.15 × 96 in.
= 110 in.

Y   Z

First floor

Joist →

Sill plate

A   Cripple wall   B

Foundation   Maximum 18 ft.

Cripple wall bracing

R602.10.9 continued   of bracing are now expressed as length in feet rather than a percentage of the braced wall line. Table R602.10.1 has been replaced by separate Tables R602.10.1.2(1) and R602.10.1.2(2) for determining the total length of bracing to resist the predominant loads from either wind or seismic forces. The eight types of bracing using isolated braced wall panels and previously represented by numbers are now known as intermittent bracing methods and have been relabeled with short abbreviations. Method WSP now represents wood structural panel bracing, replacing the method 3 designation.

R602.11
Wall Anchorage

CHANGE TYPE: Clarification

CHANGE SUMMARY: Braced wall panel connections to wood framing at interior and exterior wall locations have been consolidated in the appropriate connections provisions in Section R602.10.6. Section R602.11 now includes only those provisions related to anchorage of the braced wall line to concrete and masonry foundations.

2009 CODE: R602.11 ~~Framing and connections for Seismic Design Categories D~~$_0$~~, D~~$_1$ ~~and D~~$_2$. **Wall anchorage.** ~~The framing and connections details of buildings located in Seismic Design Categories D~~$_0$~~, D~~$_1$ ~~and D~~$_2$ ~~shall be in accordance with Sections R602.11.1 through R602.11.3.~~ Braced wall line sills shall be anchored to concrete or masonry foundations in accordance with Sections R403.1.6 and R602.11.1.

602.11.1 Wall Anchorage for All Buildings in Seismic Design Categories D$_0$, D$_1$, and D$_2$ and Townhouses in Seismic Design Category C. ~~Braced wall line sills shall be anchored to concrete or masonry foundations in accordance with Sections R403.1.6 and R602.11.~~ Plate washers, a minimum of 0.229 inch by 3 inches by 3 inches (5.8 mm by 76 mm by 76 mm) in size, shall be provided between the foundation sill plate and the nut except where approved anchor straps are used. The hole in the plate washer is permitted to be

R602.11 continues

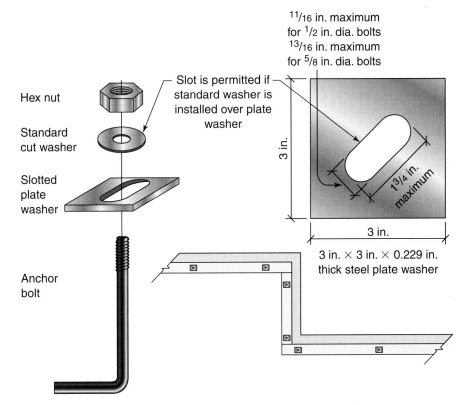

11/16 in. maximum
for 1/2 in. dia. bolts
13/16 in. maximum
for 5/8 in. dia. bolts

Slot is permitted if standard washer is installed over plate washer

Hex nut

Standard cut washer

Slotted plate washer

Anchor bolt

3 in.

1 3/4 in. maximum

3 in.

3 in. × 3 in. × 0.229 in. thick steel plate washer

Braced wall line sill anchorage to concrete or masonry foundations in all buildings in SDC D$_0$, D$_1$, and D$_2$ and townhouses in SDC C

R602.11 continued

diagonally slotted with a width of up to $^3/_{16}$ inch (5 mm) larger than the bolt diameter and a slot length not to exceed 1¾ inches (44 mm), provided a standard cut washer is placed between the plate washer and the nut.

R602.11.2 Interior Braced Wall Panel Connections. Interior braced wall lines shall be fastened to floor and roof framing in accordance with Table R602.3(1), to required foundations in accordance with Section R602.11.1, and in accordance with the following requirements:

1. Floor joists parallel to the top plate shall be toe-nailed to the top plate with at least 8d nails spaced a maximum of 6 inches (152 mm) on center.

2. Top plate laps shall be face-nailed with at least eight 16d nails on each side of the splice.

R602.11.3 R602.11.2 Stepped Foundations in Seismic Design Categories D₀, D₁, and D₂. Where stepped foundations occur, the following requirements apply:

1. Where the height of a required braced wall panel that extends from foundation to floor above varies more than 4 feet (1220 mm), the braced wall panel shall be constructed in accordance with Figure R602.11.3. In all buildings located in Seismic Design Categories D_0, D_1, and D_2, where the height of a required braced wall line that extends from foundation to floor above varies more than 4 feet (1220 mm), the braced wall line shall be constructed in accordance with the following:

2. 1. Where the lowest floor framing rests directly on a sill bolted to a foundation not less than 8 feet (2440 mm) in length along a line of bracing, the line shall be considered as braced. The double plate of the cripple stud wall beyond the segment of footing that extends to the lowest framed floor shall be spliced by extending the upper top plate a minimum of 4 feet (1219 mm) along the foundation. Anchor bolts shall be located a maximum of 1 foot and 3 feet (305 and 914 mm) from the step in the foundation. See Figure R602.11.2.

3. 2. Where cripple walls occur between the top of the foundation and the lowest floor framing, the bracing requirements for a story of Sections R602.10.9 and R602.10.9.1 shall apply.

4. 3. Where only the bottom of the foundation is stepped and the lowest floor framing rests directly on a sill bolted to the foundations, the requirements of Sections R403.1.6 and R602.11.1 shall apply.

CHANGE SIGNIFICANCE: The code no longer differentiates interior from exterior braced wall panels or braced wall lines. As part of the effort to reorganize the wall bracing provisions of Sections R602.10 and R602.11, redundant language has been removed and all provisions related to braced wall panel connections to wood framing of floor and roof/ceiling diaphragms are now located in Section R602.10.6. Section R602.11, Wall Anchorage, consolidates requirements for anchoring the sill plate of the braced wall line to a concrete or masonry foundation.

Reorganization of this section clarifies that Section R403.1.6 applies to the sill anchorage of braced wall lines for all buildings in Seismic Design Categories (SDCs) A and B, and for one- and two-family dwellings in SDC C. The anchorage provisions of Section R602.11.1 apply to all buildings in SDCs D_0, D_1, and D_2 and townhouses in SDC C. The stepped foundation provisions related to wall bracing do not apply to buildings sited in SDC A, B, or C. Changes to this section are consistent with the effort to place seismic provisions in the section where they are applicable to make the bracing provisions more user friendly and eliminate the need to thumb back and forth throughout the code to locate requirements.

R603

Steel Wall Framing

CHANGE TYPE: Modification

CHANGE SUMMARY: Section R603 has undergone extensive revision and expansion to clarify and update the prescriptive provisions for cold-formed steel light frame wall construction. These changes correlate the requirements to those in the new referenced standard AISI S230, *Standard for Cold-Formed Steel Framing—Prescriptive Method for One- and Two-Family Dwellings,* 2007 edition. The standard and the 2009 IRC expand the scope of the prescriptive methods to include three-story buildings, an increase from the previous limitation of two stories.

2009 CODE: R603.1.1 Applicability Limits. The provisions of this section shall control the construction of exterior <u>cold-formed</u> steel wall framing and interior load-bearing <u>cold-formed</u> steel wall framing for buildings not more than 60 feet (18 288 mm) long perpendicular to the joist or truss span, not more than 40 feet (12 192mm) wide parallel to the joist or truss span, and <u>less than or equal to three</u> ~~not more than two~~ stories <u>above grade plane</u> ~~in height~~. All exterior walls installed in accordance with the provisions of this section shall be considered as load-bearing walls. <u>Cold-formed</u> steel walls constructed in accordance with the provisions of this section shall be limited to sites subjected to a maximum design wind speed of 110 miles per hour (49 m/s) Exposure B or C and a maximum ground snow load of 70 pounds per square foot (3.35 kPa).

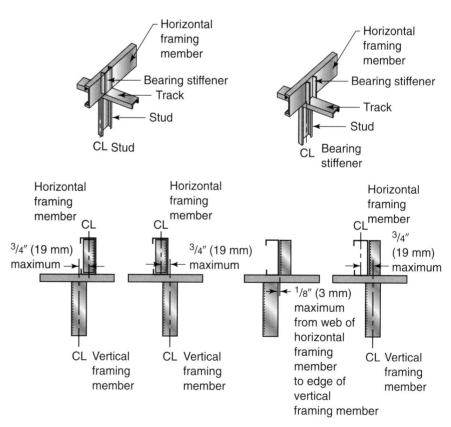

In-line framing

Because this code change deleted, added, and modified substantial portions of Section R603, the entire code change text is too extensive to be included here. Refer to the *2009 IRC Code Changes Resource Collection* for the complete text and history of code change RB168-07/08 related to Section R603.

CHANGE SIGNIFICANCE: Extensive revisions to the cold-formed steel wall framing provisions reflect the 2007 edition of AISI S230, *Standard for Cold-Formed Steel Framing—Prescriptive Method for One- and Two-Family Dwellings*. The most significant change increases the allowable height of dwellings from two to three stories above grade plane. Information to support this increase in building height has come from both testing and engineering analysis using the latest standards, including ASCE 7-05. As a result of this increase in allowable height, new tables have been added to incorporate maximum stud heights and header spans for three-story dwellings. Tables also have been revised to cover structural sheathing requirements for one-, two-, and three-story buildings.

Section R603 now includes framing details and new header tables for gable end walls intended to clarify the prescriptive provisions. All requirements concerning web holes and web hole adjustments are now consolidated in one location, clarifying that the code user has the choice to reinforce nonconforming holes, patch nonconforming holes, or design nonconforming holes with accepted engineering practice.

To clarify the application of the cold-formed steel framing provisions, many new figures have been added to the code. For example, new figures illustrate provisions for wall to foundation connections.

Other changes reflect standard practice within the cold-formed steel framing industry. For example, revised Table R603.2(2) reflects current industry standardized thickness for structural members expressed as *base steel thickness* in mils. *Reference Gage Number* is no longer used in referencing structural members and has been removed from the table.

R606.3 and R606.4

Corbeled Masonry

CHANGE TYPE: Modification

CHANGE SUMMARY: Section 606.3 has been divided into three subsections to clarify the masonry corbeling requirements. The code now specifically recognizes masonry units filled with mortar or grout as adequate for corbeling.

2009 CODE: R606.3 Corbeled Masonry.   Corbeled masonry shall be in accordance with Sections R606.3.1 through R606.3.3.

R606.3.1 Units.   Solid masonry units or masonry units filled with mortar or grout shall be used for corbeling.

R606.3.2 Corbel Projection.   ~~The maximum corbeled projection beyond the face of the wall shall not be more than one-half of the wall thickness or one-half the wythe thickness for hollow walls.~~ The maximum projection of one unit shall not exceed one-half the height of the unit or one-third the thickness at right angles to the wall. The maximum corbeled projection beyond the face of the wall shall not exceed:

1. One-half of the wall thickness for multi-wythe walls bonded by mortar or grout and wall ties or masonry headers, or
2. One-half the wythe thickness for single-wythe walls, masonry-bonded hollow walls, multi-wythe walls with open collar joints, and veneer walls.

R606.3.3 Corbeled Masonry Supporting Floor or Roof-Framing Members.   When corbeled masonry is used to support floor or roof-framing members, the top course of the corbel shall be a header course,

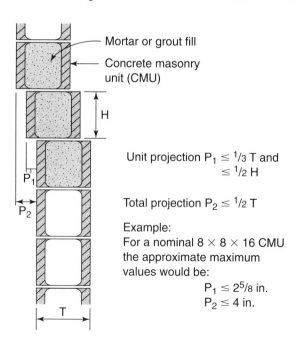

Mortar or grout fill

Concrete masonry unit (CMU)

Unit projection $P_1 \leq \frac{1}{3} T$ and $\leq \frac{1}{2} H$

Total projection $P_2 \leq \frac{1}{2} T$

Example:
For a nominal $8 \times 8 \times 16$ CMU the approximate maximum values would be:
$P_1 \leq 2\frac{5}{8}$ in.
$P_2 \leq 4$ in.

Corbeled masonry using concrete masonry units (CMU) filled with mortar or grout

or the top course bed joint shall have ties to the vertical wall. ~~The hollow space behind the corbeled masonry shall be filled with mortar or grout.~~

R606.4.2 Support at Foundation. Cavity wall or masonry veneer construction may be supported on an 8-inch (203 mm) foundation wall, provided that the 8-inch (203 mm) wall is corbeled <u>to the width of the wall system above,</u> with <u>masonry constructed of</u> solid masonry <u>units or masonry units filled with mortar or grout</u> ~~to the width of the wall system above~~. The total horizontal projection of the corbel shall not exceed 2 inches (51 mm), with individual corbels projecting not more than one-third the thickness of the unit or one-half the height of the unit. <u>The hollow space behind the corbeled masonry shall be filled with mortar or grout.</u>

CHANGE SIGNIFICANCE: Corbeling is a method of projecting successive courses of masonry out from the face of the wall below. A corbel may form a structural support or may be a decorative architectural element. The code limits the amount of projection for each course and the greatest overall projection to ensure that all loads are adequately transferred to the supporting wall below.

Previously, the IRC prescribed solid masonry units for corbeling. The change recognizes that corbelled masonry units filled with mortar or grout act the same as solid units in supporting the construction above and distribute the load as effectively as solid masonry units. Solid masonry units are not always available, whereas units filled solid with mortar or grout can be readily made on the job site as they are needed, providing more flexibility to the builder.

Section R606.3 has been reorganized and divided into three subsections to clarify the provisions and provide additional details based on the wall type. For example, the maximum corbeled projection for bonded multi-wythe walls with wall ties or masonry headers is determined by the overall wall thickness. For single-wythe walls, masonry bonded hollow walls, multi-wythe walls with open collar joints, and veneer walls, the maximum projection is based on the wythe thickness.

The requirement to fill the hollow space behind the corbel with mortar or grout has been relocated to the foundation support provisions in Section R606.4. The mortar- or grout-filled space serves to impede water entry into the foundation from the cavity or air space above. This requirement is unrelated to whether the corbel supports floor or roof-framing members and is more appropriately placed in the foundation requirements.

R606.12.2.1 and Table R606.12.2.1

Minimum Length of Masonry Walls Without Openings

CHANGE TYPE: Addition

CHANGE SUMMARY: This change adds prescriptive requirements for minimum lengths of masonry walls to provide wall bracing. The new provisions apply to above-grade masonry wall construction for townhouses located in Seismic Design Category (SDC) C and all buildings sited in SDC D_0, D_1, or D_2.

2009 CODE: <u>**R606.12.2.1 Minimum Length of Wall Without Openings.** Table R606.12.2.1 shall be used to determine the minimum required solid wall length without openings at each masonry exterior wall. The provided percentage of solid wall length shall only include those wall segments that are 3 feet (914 mm) or longer. The maximum clear distance between wall segments included in determining the solid wall length shall not exceed 18 feet (5486 mm). Shear wall segments required to meet the minimum wall length shall be in accordance with Section R606.12.2.2.3.</u>

CHANGE SIGNIFICANCE: Unlike the provisions for wood, cold-formed steel, and concrete wall systems, previous editions of the IRC did not address minimum lengths of masonry walls to resist lateral loads parallel to the wall. Section R606.12.2.1 and Table R606.12.2.1 add prescriptive masonry wall bracing requirements for high Seismic Design Categories (SDCs). As established in the charging statement of Section R606.12, these new provisions apply to townhouses sited in SDC C and all buildings located in SDC D0, D1, or D2. The minimum solid wall length along exterior masonry wall lines was developed in part to correlate to the minimum length requirements for insulated

Length of solid wall segments	Minimum solid wall length	Seismic Design Category (SDC)
A + B + C ≥	20% of L	Townhouse SDC C
	25% of L	All buildings in SDC D_0, D_1
	30% of L	All buildings in SDC D_2

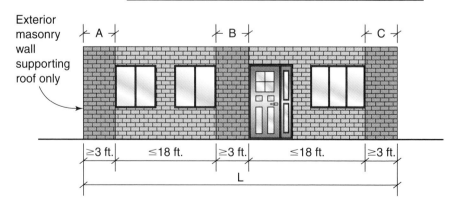

Minimum length of masonry walls without openings in Seismic Design Categories C, D_0, D_1, and D_2

concrete form (ICF) walls. *Solid walls* in this case refer to full-height wall segments without openings, and the code does not intend to require the walls to be grouted solid. The minimum length requirements are expressed as a percentage of the overall wall length. Only solid wall segments 3 feet in length and greater contribute to the minimum percentage requirements.

TABLE R606.12.2.1 **Minimum Solid Wall Length Along Exterior Wall Lines**

| Seismic Design Category | MINIMUM SOLID WALL LENGTH (percent)[a] | | |
	One-Story or Top Story of Two Story	Wall Supporting Light-Framed Second Story and Roof	Wall Supporting Masonry Second Story and Roof
Townhouses in SDC C	20	25	35
D_0 or D_1	25	NP	NP
D_2	30	NP	NP

NP = Not permitted, except with design in accordance with the International Building Code.

a. For all walls, the minimum required length of solid walls shall be based on the table percent multiplied by the dimension, parallel to the wall direction under consideration, of a rectangle inscribing the overall building plan.

R611

Exterior Concrete Wall Construction

CHANGE TYPE: Modification

CHANGE SUMMARY: Section R611 has been completely revised to reflect the provisions of the new referenced Portland Cement Association standard PCA 100 *Prescriptive Design of Exterior Concrete Walls for One- and Two-Family Dwellings.* Conventionally formed above-ground concrete wall provisions have been integrated with the insulated concrete form (ICF) wall requirements.

2009 CODE: **Section R611 ~~Insulating~~ <u>Exterior</u> Concrete ~~Form~~ Wall Construction**

R611.1 General. ~~Insulating~~ <u>Exterior</u> Concrete ~~Form (IFC)~~ walls shall be designed and constructed in accordance with the provisions of this section or in accordance with the provisions of <u>PCA 100</u> or ACI 318. When <u>PCA 100,</u> ACI 318 or the provisions of this section are used to design ~~insulating~~ concrete ~~form~~ walls, project drawings, typical details, and specifications are not required to bear the seal of the architect or engineer responsible for design, unless otherwise required by the state law of the jurisdiction having authority.

R611.2 Applicability Limits. The provisions of this section shall apply to the construction of ~~insulating~~ exterior concrete ~~form~~ walls for

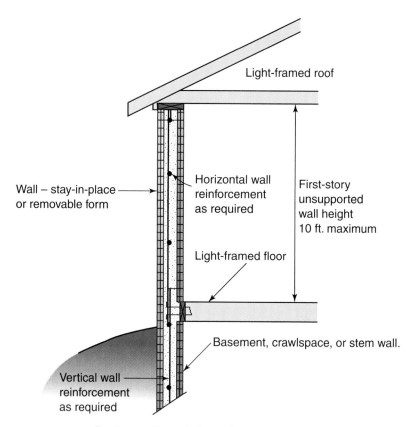

Section cut through flat wall or vertical
core of a waffle- or screen-grid wall

Exterior concrete wall construction

buildings not greater than 60 feet (18 288 mm) in plan dimensions, ~~and~~ floors <u>with clear spans</u> not greater than 32 feet (9754 mm) ~~or~~ <u>and</u> roofs <u>with clear spans</u> not greater than 40 feet (12 192 mm) ~~in clear span~~. Buildings shall not exceed <u>35 feet in mean roof height or</u> two stories in height above grade. <u>Floor/ceiling dead loads shall not exceed 10 pounds per square foot (479 Pa), roof/ceiling dead loads shall not exceed 15 pounds per square foot (720 Pa), and attic live loads shall not exceed 20 pounds per square foot (958 Pa). Roof overhangs shall not exceed 2 feet (610 mm) of horizontal projection beyond the exterior wall, and the dead load of the overhangs shall not exceed 8 pounds per square foot (383 Pa).</u> ~~ICF walls shall comply with the requirements in Table R611.2.~~

Walls constructed in accordance with the provisions of this section shall be limited to buildings subjected to a maximum design wind speed of ~~150 miles per hour (67 m/s),~~ <u>130 miles per hour (58 m/s) Exposure B, 110 miles per hour (49 m/s) Exposure C, and 100 miles per hour (45 m/s) Exposure D</u> ~~and~~. <u>Walls constructed in accordance with the provisions of this section shall be limited to detached one- and two-family dwellings and townhouses assigned to Seismic Design Category A or B, and detached one- and two-family dwellings assigned to Seismic Design Category C, D_0, D_1 and D_2.</u> ~~The provisions of this section shall not apply to the construction of ICF walls for buildings or portions of buildings considered irregular as defined in Section R301.2.2.2.2.~~

<u>Buildings that are not within the scope of this section shall be designed in accordance with PCA 100 or ACI 318.</u> ~~For townhouses in Seismic Design Category C and all buildings in Seismic Design Category D_0, D_1 or D_2, the provisions of this section shall apply only to buildings meeting the following requirements.~~

~~1. Rectangular buildings with a maximum building aspect ratio of 2:1. The building aspect ratio shall be determined by dividing the longest dimension of the building by the shortest dimension of the building.~~

~~2. Walls are aligned vertically with the walls below.~~

~~3. Cantilever and setback construction shall not be permitted.~~

~~4. The weight of interior and exterior finishes applied to ICF walls shall not exceed 8 psf (380 Pa).~~

~~5. The gable portion of ICF walls shall be constructed of light-frame construction.~~

~~Section R612~~
~~Conventionally Formed Concrete Wall Construction~~
~~**R612.1 General.** Conventionally formed concrete walls with flat surfaces shall be designed and constructed in accordance with the provisions of Section R611 for Flat ICF walls or in accordance with the provisions of ACI 318.~~

Because this code change deleted, added, and modified substantial portions of Section R611, the entire code change text is too extensive to be included here. Refer to Code Change RB171-07/08 in the *2009 IRC Code Changes Resource Collection* for the complete text and history of the code change.

R611 continues

R611 continued

CHANGE SIGNIFICANCE: Previously, the above-ground concrete wall provisions of Section R612 referenced the design and construction requirements of Section R611 for flat insulated concrete form (ICF) walls or ACI 318. With this change, the conventionally formed concrete wall provisions are merged and correlated with those for ICF walls in the substantially revised provisions of Section R611. Section R612 has been deleted. Revision of the prescriptive provisions for above-grade concrete walls reflect the provisions of a new consensus standard developed by the Portland Cement Association, PCA 100 *Prescriptive Design of Exterior Concrete Walls for One- and Two-Family Dwellings.* This new standard replaces PCA publication *Prescriptive Method for Insulating Concrete Forms in Residential Construction,* which served as the basis for most of the ICF concrete wall provisions in Section R611 of the 2006 IRC. As alternatives to the prescriptive provisions or where the design exceeds the scope of the IRC, the code references PCA 100 and ACI 318, *Building Code Requirements for Structural Concrete.*

The applicability limits of Section R611 are generally consistent with the limitations of PCA 100 for building plan dimensions, height, projections, and dead loads. The referenced standard does contain provisions for construction in higher design wind speed regions, up to 150 mph in Exposure Category D. The prescriptive provisions of the 2009 IRC for above-grade exterior concrete walls are now limited to buildings where the maximum design wind speed is 130 mph in Exposure B, 110 mph in Exposure C, and 100 mph in Exposure D. Buildings located in higher design wind speed regions must be designed in accordance with PCA 100 or ACI 318. In addition, the IRC no longer includes prescriptive provisions for above-ground concrete walls in high Seismic Design Categories (SDCs). Section R611 is limited to detached one- and two-family dwellings and townhouses located in SDC A or B, and detached one- and two-family dwellings in SDC C. The limited provisions for concrete walls of buildings in high seismic areas that appeared in the 2006 IRC have been deleted but still appear in PCA 100. Townhouses in SDC C and all buildings in SDC D must now meet the requirements of PCA 100 or be designed in accordance with ACI 318.

Revised figures and tables for constructing flat, waffle-grid, and screen-grid ICF wall systems appear in Section R611.5. Additional provisions are included for constructing these concrete walls based on concrete, aggregate, and steel reinforcement materials used. New provisions govern the location, cover, and continuity of the reinforcement and installation of construction joints.

The prescriptive technical provisions for exterior concrete walls in Section R611.6 through R611.8 have been replaced entirely and reflect changes made to ACI 318 and ASCE 7 since the provisions in the existing code were developed. The new provisions cover horizontal and vertical reinforcement, reinforcement and shear wall (solid wall) requirements for wind loads, and revised reinforcement requirements around openings and lintels over openings. Section R611.9 has also been replaced with revised details for connecting wood and cold-formed steel framing assemblies to exterior concrete walls.

R612.2
Window Sills

CHANGE TYPE:　Modification

CHANGE SUMMARY:　Changes to Sections R612.2 through R612.4 clarify the child fall prevention alternatives to the minimum window sill height. In the first alternative, *window fall prevention device* replaces the term *guard* as the barrier installed at operable windows with sills below 24 inches. In the second option, the code details the performance criteria for opening limiting devices, including provisions for emergency escape and rescue openings.

2009 CODE:　R612.2 Window Sills.　In dwelling units, where the opening of an operable window is located more than 72 inches (1829 mm) above the finished grade or surface below, the lowest part of the clear opening of the window shall be a minimum of 24 inches (610 mm) above the finished floor of the room in which the window is located. ~~Glazing between the floor and 24 inches (610 mm) shall be fixed or have openings through which a 4-inch-diameter (102 mm) sphere cannot pass.~~ <u>Operable sections of windows shall not permit openings that allow passage of a 4-inch-diameter sphere where such openings are located within 24 inches of the finished floor.</u>

Exceptions:

1. Windows whose openings will not allow a 4-inch-diameter (102 mm) sphere to pass through the opening when the opening is in its largest opened position.

2. Openings that are provided with window <u>fall prevention devices</u> ~~guards~~ that comply with <u>R612.3</u> ~~ASTM F 2006 or F 2090~~.

Window opening limiting devices and fall prevention devices must be approved for emergency escape and rescue provisions

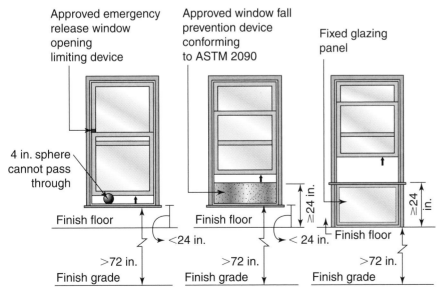

Approved emergency release window opening limiting device

Approved window fall prevention device conforming to ASTM 2090

Fixed glazing panel

4 in. sphere cannot pass through

Finish floor

<24 in.

>72 in.

Finish grade

Finish floor

< 24 in.

≥24 in.

>72 in.

Finish grade

Finish floor

≥24 in.

>72 in.

Finish grade

Window sill height

R612.2 continues

R612.2 continued

3. Openings that are provided with fall prevention devices that comply with ASTM F 2090.

4. Windows that are provided with opening limiting devices that comply with Section R612.4.

R612.3 Window Fall Prevention Devices. Window fall prevention devices and window guards, where provided, shall comply with the requirements of ASTM F 2090.

R612.4 Window Opening Limiting Devices. When required elsewhere in this code, window opening limiting devices shall comply with the provisions of this section.

R612.4.1 General Requirements. Window opening limiting devices shall be self-acting and shall be positioned so as to prohibit the free passage of a 4-inch (102 mm)-diameter rigid sphere through the window opening when the window opening limiting device is installed in accordance with the manufacturer's instructions.

R612.4.2 Operation for Emergency Escape. Window opening limiting devices shall be designed with release mechanisms to allow for emergency escape through the window opening without the need for keys, tools or special knowledge. Window opening limiting devices shall comply with all of the following:

1. Release of the window opening limiting device shall require no more than 15 lbs (66 N) of force.

2. The window opening limiting device release mechanism shall operate properly in all types of weather.

3. Window opening limiting devices shall have their release mechanisms clearly identified for proper use in an emergency.

4. The window opening limiting device shall not reduce the minimum net clear opening area of the window unit below what is required by Section R310.1.1 of the code.

CHANGE SIGNIFICANCE: The 2009 IRC maintains the requirement for a 24-inch minimum window sill height when the opening is more than 72 inches above grade, a provision that intends to prevent small children from falling out of open windows. This change clarifies the application of this requirement and updates and expands the information on the alternatives for fall prevention when the sill is lower than 24 inches above the floor—installing a barrier or limiting the dimensions of the window opening.

The first exception remains unchanged and permits installation of a window that is manufactured such that, when opened, it does not allow a 4-inch-diameter sphere to pass through. This option is not permitted in a location requiring an emergency escape and rescue opening. Window opening limiting devices, on the other hand, are now specifically approved for emergency escape and rescue windows when they meet the new requirements. An opening limiting device installed on any window must have an emergency release device that is clearly identified and that operates without the need for a key, tool, or special knowledge. These operation criteria match the language in the provisions for emergency escape and rescue openings. The code also limits the opening force of the release mechanism. There presently is no approved standard for window opening limiting or control devices.

The other option for windows with sills lower than 24 inches above the floor is to provide a barrier at the window opening that does not permit passage of a 4-inch-diameter sphere. To avoid confusion with the 36-inch-high guard requirements in Section R310 or limit the design of the protective device, and for consistency with the referenced standard, the term *guard* has been replaced with *window fall prevention device*. The code references ASTM F 2090, *Window Fall Prevention Devices with Emergency Escape (Egress) Release Mechanisms* for the device requirements. The standard requires window fall prevention devices to be constructed such that a 4-inch-diameter sphere cannot pass through. The previous referenced standard, ASTM F 2006 *Standard Safety Specification for Window Fall Prevention Devices for Non-Emergency Escape (Egress) and Rescue (Ingress) Windows*, is no longer approved for fall protection under these code provisions. Window fall prevention devices installed on any window must conform to ASTM F 2090, thereby complying with the operation provisions for emergency escape and rescue openings in Section R310.

R613

Structural Insulated Panel Wall Construction

CHANGE TYPE: Addition

CHANGE SUMMARY: Prescriptive provisions for structural insulated panel (SIP) wall construction have been added to the code in a new Section R613. Application of these provisions is limited to one- and two-story buildings not greater than 40 feet wide by 60 feet long with 10-foot wall heights and sited in Seismic Design Categories A, B, and C. Maximum design wind speed is 130 mph in Exposure C and maximum snow load is 70 psf.

2009 CODE:
Section R613
Structural Insulated Panel Wall Construction

R613.1 General. Structural insulated panel (SIP) walls shall be designed in accordance with the provisions of this section. When the provisions of this section are used to design structural insulated panel walls, project drawings, typical details, and specifications are not required to bear the seal of the architect or engineer responsible for design, unless otherwise required by the state law of the jurisdiction having authority.

R613.2 Applicability Limits. The provisions of this section shall control the construction of exterior structural insulated panel walls and interior load-bearing structural insulated panel walls for buildings not greater than 60 feet (18 288 mm) in length perpendicular to the joist or truss span, not greater than 40 feet (10 973 mm) in width parallel to the joist or truss span and not greater than two stories in height, with each wall not greater than 10 feet (3048 mm) high. All exterior walls installed in accordance with the provisions of this section shall be considered as load-bearing walls. Structural insulated panel walls constructed in accordance with the provisions of this section shall be limited to sites subjected to a maximum design wind speed of 130 miles per hour, Exposure A, B, or C, and a maximum ground snow load of 70 pounds per foot (3.35 kN/m^2), and Seismic Design Categories A, B, and C.

R613.4 SIP Wall Panels. SIP panels shall comply with Figure R613.4 and shall have minimum panel thickness in accordance with Tables R613.5(1) and R613.5(2) for above-grade walls. All SIPs shall be identified by grade mark or certificate of inspection issued by an approved agency.

Because this code change added a new Section R613 with substantial information, the entire code change text is too extensive to be included here. Refer to the *2009 IRC Code Changes Resource Collection* for the complete text and history of the code change.

CHANGE SIGNIFICANCE: With this change, the IRC includes prescriptive provisions recognizing structural insulated panels (SIPs) for exterior and interior bearing wall construction. An SIP consists of a light-weight plastic foam insulation core securely laminated between

TABLE R613.5(2) **Minimum Thickness for SIP Walls Supporting SIP or Light-Frame One Story and Roof (inches)**

Wind Speed (3-sec. gust)		Snow Load (psf)	Building Width (ft)															
			24			28			32			36			40			
Exp. A/B	Exp. C		Wall Height (feet)			Wall Height (feet)			Wall Height (feet)			Wall Height (feet)			Wall Height (feet)			
			8	9	10	8	9	10	8	9	10	8	9	10	8	9	10
85		20	4.5	4.5	4.5	4.5	4.5	4.5	4.5	4.5	4.5	4.5	4.5	4.5	4.5	4.5	4.5
		30	4.5	4.5	4.5	4.5	4.5	4.5	4.5	4.5	4.5	4.5	4.5	4.5	4.5	4.5	4.5
		50	4.5	4.5	4.5	4.5	4.5	4.5	4.5	4.5	4.5	4.5	4.5	4.5	4.5	4.5	4.5
		70	4.5	4.5	4.5	4.5	4.5	4.5	4.5	4.5	4.5	4.5	4.5	6.5	6.5	6.5	6.5
100	85	20	4.5	4.5	4.5	4.5	4.5	4.5	4.5	4.5	4.5	4.5	4.5	4.5	4.5	4.5	4.5
		30	4.5	4.5	4.5	4.5	4.5	4.5	4.5	4.5	4.5	4.5	4.5	4.5	4.5	4.5	6.5
		50	4.5	4.5	4.5	4.5	4.5	4.5	4.5	4.5	4.5	4.5	4.5	6.5	4.5	6.5	6.5
		70	4.5	4.5	4.5	4.5	4.5	4.5	4.5	4.5	6.5	6.5	6.5	6.5	6.5	N/A	N/A
110	100	20	4.5	4.5	4.5	4.5	4.5	4.5	4.5	4.5	4.5	4.5	4.5	4.5	4.5	4.5	6.5
		30	4.5	4.5	4.5	4.5	4.5	4.5	4.5	4.5	4.5	4.5	4.5	6.5	4.5	6.5	6.5
		50	4.5	4.5	4.5	4.5	4.5	4.5	4.5	4.5	6.5	4.5	6.5	6.5	6.5	6.5	N/A
		70	4.5	4.5	4.5	4.5	4.5	6.5	6.5	6.5	N/A	6.5	N/A	N/A	N/A	N/A	N/A
120	110	20	4.5	4.5	4.5	4.5	4.5	4.5	4.5	4.5	6.5	4.5	4.5	6.5	4.5	6.5	N/A
		30	4.5	4.5	4.5	4.5	4.5	6.5	4.5	4.5	6.5	4.5	6.5	N/A	6.5	6.5	N/A
		50	4.5	4.5	6.5	4.5	4.5	6.5	4.5	6.5	N/A	6.5	N/A	N/A	N/A	N/A	N/A
		70	4.5	4.5	6.5	4.5	6.5	N/A	6.5	N/A	N/A	N/A	N/A	N/A	N/A	N/A	N/A
130	120	20	4.5	4.5	6.5	4.5	4.5	6.5	4.5	6.5	N/A	4.5	6.5	N/A	6.5	N/A	N/A
		30	4.5	4.5	6.5	4.5	4.5	N/A	4.5	6.5	N/A	6.5	N/A	N/A	6.5	N/A	N/A
		50	4.5	6.5	N/A	4.5	6.5	N/A	6.5	N/A	N/A	N/A	N/A	N/A	N/A	N/A	N/A
		70	4.5	6.5	N/A	6.5	N/A	N/A	N/A	N/A	N/A	N/A	N/A	N/A	N/A	N/A	N/A
	130	20	6.5	N/A	6.5	N/A	N/A	N/A	N/A	N/A	N/A	N/A	N/A	N/A	N/A	N/A	N/A
		30	6.5	N/A	N/A	N/A	N/A	N/A	N/A	N/A	N/A	N/A	N/A	N/A	N/A	N/A	N/A
		50	N/A	N/A	N/A	N/A	N/A	N/A	N/A	N/A	N/A	N/A	N/A	N/A	N/A	N/A	N/A
		70	N/A	N/A	N/A	N/A	N/A	N/A	N/A	N/A	N/A	N/A	N/A	N/A	N/A	N/A	N/A

For SI: 1 inch = 25.4 mm; 1 foot = 304.8 mm; 1 pound per square foot = 0.0479 kPa.
Maximum deflection criteria: L/240.
Maximum roof dead load: 10 psf.
Maximum roof live load: 70 psf.
Maximum ceiling dead load: 5 psf.
Maximum ceiling live load: 20 psf.
Maximum second floor live load: 30 psf.
Maximum second floor dead load: 10 psf.
Maximum second floor dead load from walls: 10 psf.
Maximum first floor live load: 40 psf.
Maximum first floor dead load: 10 psf.
Wind loads based on Table R301.2 (2).
N/A indicates not applicable.

two wood structural panel facings not less than $^7/_{16}$ inch thick. The structural characteristics of SIPs are similar to those for a steel I-beam. The facings act like the flanges of an I-beam, and the rigid core provides the web of the I-beam configuration. This composite assembly produces stiffness, strength, and predictable performance. The new

R613 continues

R613 continued

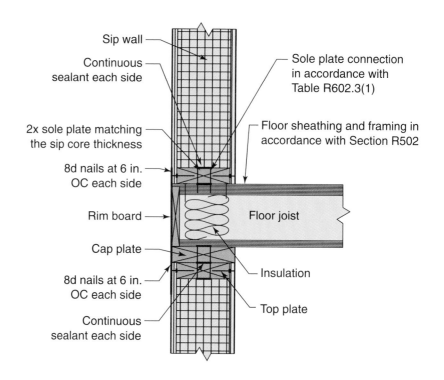

Structural insulated panel (SIP) wall detail

provisions are based on testing using industry-developed minimum properties for panels, adhesives, and foam density. Tests included axial, shear, and transverse loads, all conducted in accordance with recognized test methods in developing panel capacities.

Similar to the prescriptive provisions for cold-formed steel framing, SIP wall construction in accordance with Section R613 is limited to buildings not greater than 40 feet wide by 60 feet long with 10-foot wall heights. Construction using these prescriptive provisions is further limited to locations with a maximum design wind speed of 130 mph

Exposure A, B, or C, a maximum ground snow load of 70, and Seismic Design Categories A, B, and C. While the 2009 edition of the IRC permits conventional wood and cold-formed steel framed buildings up to three stories in height, the prescriptive provisions of Section R613 limit SIP construction to not more than two stories.

Section R613 contains prescriptive tables, material specifications, bracing information, and construction details similar to those found in the wood and cold-formed steel framing and concrete wall sections of the code. The minimum thickness of SIP for a particular application is determined in accordance with Tables R613.5(1) and R613.5(2) based on building width, design wind speed, snow load, and elements being supported. In addition, certain combinations of these design criteria fall outside the scope of the prescriptive provisions. For example, Section R613 does not permit SIP construction for 10-foot walls supporting one story and a roof for a 40-foot-wide building exposed to a 110 mph wind speed in exposure C. All SIPs must bear a grade mark or certificate of inspection issued by an approved agency.

SIP panels are connected at their vertical edges with splines that fit behind the facings of the panels. The top and bottom of the SIP panels are recessed to receive a 2-inch nominal thick plate that matches the width of the plastic foam core. A cap plate is required above the first top plate and matches the overall width of the panel to ensure that concentrated loads are carried by the facings of the SIP panel and not just by the foam core.

R703 and Table R703.4

Weather-Resistant Exterior Covering

CHANGE TYPE: Modification

CHANGE SUMMARY: Performance requirements for wind resistance have been added to the water resistance provisions of exterior wall covering systems in Section R703.1. Changes to Table R703.4 clarify the water-resistive barrier requirements for various wall covering and cladding systems and update the fastening requirements to reflect current industry practices.

SECTION R703 EXTERIOR COVERING

R703.1 General. Exterior walls shall provide the building with a weather-resistant exterior wall envelope. The exterior wall envelope shall include flashing as described in Section R703.8.

R703.1.1 Water Resistance. The exterior wall envelope shall be designed and constructed in a manner that prevents the accumula-

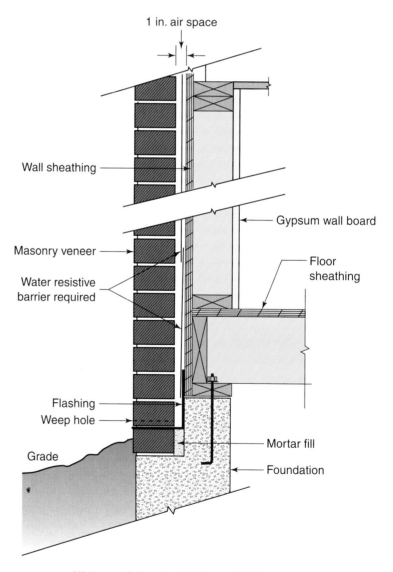

Water-resistive barrier is required over wall sheathing behind brick veneer

TABLE R703.4 Weather-Resistant Siding Attachment and Minimum Thickness

Siding Material		Nominal Thick-ness[a] (inches)	Joint Treat-ment	Water-Resistive Barrier Required	Type of Supports for the Siding Material and Fasteners[b,c,d]					Number or Spacing of Fasteners
					Wood or Wood Structural Panel Sheathing	Fiberboard Sheathing into Stud	Gypsum Sheathing into Stud	Foam Plastic Sheath-ing into Stud	Direct to Studs	
Horizontal aluminum[e]	Without insulation	0.019[f]	Lap	Yes	0.120 nail 1 2″ long	0.120 nail 2″ long	0.120 nail 2″ long	0.120 nail[y v]	Not allowed	Same as stud spacing
		0.024	Lap	Yes	0.120 nail 1 2″ long	0.120 nail 2″ long	0.120 nail 2″ long	0.120 nail[y v]	Not allowed	
	With insulation	0.019	Lap	Yes	0.120 nail 1 2″ long	0.120 nail 22″ long	0.120 nail 22″ long	0.120 nail[y v]	0.120 nail 1 2″ long	
Anchored veneer: brick, concrete, masonry, or stone ~~Brick veneer~~[z] ~~Concrete masonry veneer~~[z]		2	Section R703	Yes ~~(Note l)~~	See Section R703 and Figure R703.7[g]					
Adhered veneer: concrete, stone, or masonry[w]			Section R703	Yes Note w	See Section R703.6.1[g] or in accordance with the manufacturer's instructions.					
Hardboard[k] panel siding: vertical		7/16	—	Yes	Note ~~n~~ m	Note ~~n~~ m	Note ~~n~~ m	Note ~~n~~ m	Note ~~n~~ m	6″ panel edges 12″ inter. sup.[ne]
Hardboard[k] lap siding: horizontal		7/16	Note ~~p~~ q	Yes	Note ~~n~~ o	Note ~~n~~ o	Note o	Note ~~n~~ o	Note ~~n~~ o	Same as stud spacing 2 per bearing
Steel[h]		29 ga.	Lap	Yes	0.113 nail 1: Staple-1:	0.113 nail 2: Staple-22	0.013 nail 22 Staple-23	0.113 nail[vy] Staple[vy]	Not allowed	Same as stud spacing
~~Stone veneer~~		~~2~~	~~Section R703~~	~~Yes (Note l)~~	~~See Section R703 and Figure R703.7~~[g]					
Particleboard panels		3/8- 1/2	—	Yes	6d box (2″ × 0.099″) nail	6d box (2″ × 0.099″) nail	6d box (2″ × 0.099″) nail	box nail[vy]	6d box (2″ × 0.099″) nail, 3/8 not allowed	6″ panel edge 12″ inter. sup.
		5/8	—	yes	6d box (2″ × 0.099″) nail	8d box (2 2″ × 0.113″) nail	8d box (2 2″ × 0.113″) nail	box nail[vy]	6d box (2″ × 0.099″) nail	
Wood structural ~~Plywood~~ panel-siding[i] (exterior grade)		3/8- 1/2	Note p	Yes	0.099 nail-2″	0.113 nail-2½″	~~0.099~~ 0.113 nail-2½	0.113 nail[vy]	0.099 nail-2″	6″ panel edges, 12″ inter. sup.
Wood structural panel lap siding		3/8-1/2	Note p Note x	Yes	0.099 Nail-2″	0.113 Nail-2-½″	0.113 Nail-2-½″	0.113 nail[x]	0.099 Nail-2″	8″ along bottom edge

R703 and Table R703.4 continues

R703 and Table R703.4 continued

TABLE R703.4　Weather-Resistant Siding Attachment and Minimum Thickness

Siding Material	Nominal Thickness[a] (inches)	Joint Treatment	Water-Resistive Barrier Required	Type of Supports for the Siding Material and Fasteners[b,c,d]					Number or Spacing of Fasteners
				Wood or Wood Structural Panel Sheathing	Fiberboard Sheathing into Stud	Gypsum Sheathing into Stud	Foam Plastic Sheathing into Stud	Direct to Studs	
Vinyl siding[l̶m]	0.035	Lap	Yes	0.120 nail 1½″ ~~Staple-1¾″~~ (shank) with a .313 head or 16 gauge staple with to ½-in. crown[y,z]	0.120 nail ~~2″ Staple-2½″~~ (shank) with a .313 head or 16 gauge staple with ⅜ to ½-in. crown[y]	0.120 nail ~~2″ Staple-2½″~~ (shank) with a .313 head or 16 gauge staple with ⅜ to ½-in. crown[y]	0.120 nail ~~y~~ Staple[y] (shank) with a .313 head per Section R703.11.2[y]	Not allowed	16 inches on center or as specified by the manufacturer instructions or test report ~~Same as stud spacing~~
Wood[j] Rustic, drop	⅜ Min	Lap	Yes	Fastener penetration into stud-1″				0.113 nail-2½″ Staple-2″	Face nailing up to 6″ widths, 1 nail per bearing; 8″ widths and over 2 nail per bearing
Shiplap	¹⁹⁄₃₂ Average	Lap	Yes						
Bevel	⁷⁄₁₆								
Butt tip	³⁄₁₆	Lap	Yes						
Fiber cement panel siding[q̶r]	⁵⁄₁₆	Note q̲ ~~s~~	Yes Note u̲ ~~x~~	6d common corrosion resistant nail[rt]	6d common corrosion resistant nail[rt]	6d common corrosion resistant nail[rt]	6d common corrosion resistant (12″ × 0.113″) nail[rt,vy]	4d common corrosion resistant nail[rt]	6″ oc on edges, 12″ oc on intermed. studs
Fiber cement lap siding[s̶ r̶]	⁵⁄₁₆	Note s̲ ~~v~~	Yes Note u̲ ~~x~~	6d common corrosion resistant nail[r̲ t]	6d common corrosion resistant nail[r̲ t]	6d common corrosion resistant nail[r̲ t]	6d common corrosion resistant (12″ × 0.113″) nail[r̲ t,v̲ y]	6d common corrosion resistant nail or 11 gage roofing nail[r̲ w]	Note t̲ ~~w~~

For SI: 1 inch = 25.4 mm.

a.–i. : (No change to text.)

j. Wood board sidings applied vertically shall be nailed to horizontal nailing strips or blocking set 24 inches on center. Nails shall penetrate 1½ inches into studs, studs and wood sheathing combined, or blocking. ~~A weather-resistive membrane shall be installed weatherboard fashion under the vertical siding unless the siding boards are lapped or battens are used.~~

k. Hardboard siding shall comply with AHA A135.6.

~~l. For Masonry veneer, a weather-resistant sheathing paper is not required over a sheathing that performs as a weather-resistive barrier when a 1-inch air space is provided between the veneer and the sheathing. When the 1-inch space is filled with mortar, a weather-resistant sheathing paper is required over studs or sheathing.~~

~~m.~~ l̲. Vinyl siding shall comply with ASTM D 3679.

~~n.~~ m̲. Minimum shank diameter of 0.092 inch, minimum head diameter of 0.025 inch, and nail length must accommodate sheathing and penetrate framing 1½ inches.

~~o.~~ n̲. When used to resist shear forces, the spacing must be 4 inches at panel edges and 8 inches on interior supports.

p. o. Minimum shank diameter of 0.099 inch, minimum head diameter of 0.240 inch, and nail length must accommodate sheathing and penetrate framing 1½ inches.

q. p. Vertical end joints shall occur at studs and shall be covered with a joint cover or shall be caulked.

r. Fiber cement siding shall comply with the requirements of ASTM C 1186.

s. q. See R703.10.1.

t. r. Fasteners shall comply with the nominal dimensions in ASTM F1667. Minimum 0.102" smooth shank, 0.255" round head.

u. Minimum 0.099" smooth shank, 0.250" round head.

v. s. See R703.10.2.

w. t. Face nailing: one 6d common nail through the overlapping planks at each stud. Concealed nailing: one 11 gage 1½ inch long galv. roofing nail through the top edge of each plank at each stud. 2 nails at each stud. Concealed nailing: one 11 gage 1½ galv. roofing nail (0.371² head diameter, 0.120² shank) or 6d galv. box nail at each stud.

x. u. See R703.2 Exceptions.

y. v. Minimum nail length must accommodate sheathing and penetrate framing 1½ inches.

z. w. Adhered masonry veneer shall comply with the requirements of Section R703.6.3 and shall comply with the requirements in Sections 6.1 and 6.3 of ACI 530/ASCE 5/TMS-402.

x. Vertical joints, if staggered, shall be permitted to be away from studs if applied over wood structural panel sheathing.

y. Minimum fastener length must accommodate sheathing and penetrate framing .75 inch or in accordance with the manufacturer's installation instructions.

z. Where approved by the manufacturer's instructions or test report, siding shall be permitted to be installed with fasteners penetrating not less than .75 inch through wood or wood structural sheathing with or without penetration into the framing.

tion of water within the wall assembly by providing a water-resistant barrier behind the exterior veneer as required by Section R703.2 and a means of draining water that enters the assembly to the exterior. Protection against condensation in the exterior wall assembly shall be provided in accordance with Section R601.3 of this code.

Exceptions: (No change to text.)

703.1.2 Wind Resistance. Wall coverings, backing materials, and their attachments shall be capable of resisting wind loads in accordance with Tables R301.2(2) and R301.2(3). Wind pressure resistance of the siding and backing materials shall be determined by ASTM E 330 or other applicable standard test methods. Where wind pressure resistance is determined by design analysis, data from approved design standards and analysis conforming with generally accepted engineering practice shall be used to evaluate the siding and backing material and its fastening. All applicable failure modes including bending rupture of siding, fastener withdrawal, and fastener head pull-through shall be considered in the testing or design analysis. Where the wall covering and the backing material resist wind load as an assembly, the design capacity of the assembly shall be permitted.

R703.3.2 Horizontal Siding. Horizontal lap siding shall be installed in accordance with the manufacturer's recommendations. Where there are no recommendations, the siding shall be lapped a minimum of 1 inch (25 mm), or ½ inch (13 mm) if rabbeted, and shall have the ends caulked, covered with a batten, or sealed and installed over a strip of flashing.

CHANGE SIGNIFICANCE: The provisions for weather resistance in Section R703.1 have been broken into two subsections to recognize both water and wind resistance for exterior wall covering systems. The water-resistance requirements in this section remain unchanged. Section R703.1.2 provides a basis for testing and analysis of wind pressure resistance of all exterior covering systems, and references the component

R703 and Table R703.4 continues

R703 and Table R703.4 continued

and cladding wind load requirements of Tables R301.2(2) and R301.2(3). This change clearly highlights and distinguishes the key performance requirements for both water and wind resistance in the appropriate location for addressing weather resistance of the exterior wall envelope.

The manufacturers of siding materials often specify minimum lap requirements that may differ from the provisions of the 2006 IRC. The change to Section R703.3.2 requires lap siding to be installed as recommended by the manufacturer. The prescriptive lap dimensions still apply in the absence of recommendations.

The omission of a water-resistive barrier behind masonry veneer with a 1-inch air space is no longer permitted by Table R703.4. This change was prompted by observations of poor building performance when the water-resistive barrier was not installed. Mortar squeeze-out and mortar falling to the bottom of the 1-inch air space often bridges the gap, causing a transfer of moisture to the sheathing surface. In addition, masonry veneer may not be installed on upper portions of a wall or on upper stories. Installation of sheathing paper or approved house wrap behind masonry veneer ensures a continuation of the water-resistive barrier intended to drain water behind the cladding material above. Proponents of the change maintain that the intent of the code is to provide a double layer of weather protection between the outside environment and the unprotected framework of the wall. Stone and masonry veneer, both very porous materials, do not form an adequate weather-resistive barrier, and the 1-inch air space is not thought to be equivalent to a water-resistive barrier.

In Table R703.4, the line item for *stone veneer* has been deleted, the row for brick and masonry veneer has been changed to *anchored veneer,* and a new line has been added for *adhered veneer. Anchored veneer* now includes brick, concrete, masonry, and stone that is secured to the structure with the code-prescribed metal ties. Lath attachments for adhered veneer must comply with Section R703.6.1 or the manufacturer's requirements. These changes improve consistency and understanding of the respective provisions for installation of the various veneer materials, whether they are anchored or adhered to the structure.

Other changes to Table R703.4 include fastening requirements for wood structural panel siding and vinyl siding to recognize current industry and manufacturer's recommendations and test reports. Vinyl siding installed over foam plastic sheathing must be fastened with approved nails in accordance with new Section R703.11.2. Staples are no longer permitted for this installation.

CHANGE TYPE: Modification

R703.7.3
Lintels

CHANGE SUMMARY: Steel lintels supporting masonry veneer above openings now require a shop coat of rust-inhibitive primer or other protection against corrosion. The 2009 IRC also provides an alternative prescriptive method for supporting veneer above openings measuring up to 18 feet 3 inches in length using a combination of a steel angle and masonry with horizontal reinforcing.

2009 CODE: **R703.7.3 Lintels.** Masonry veneer shall not support any vertical load other than the dead load of the veneer above. Veneer above openings shall be supported on lintels of noncombustible materials ~~and the allowable span shall not exceed the values set forth in Table R703.7.3.3~~. The lintels shall have a length of bearing not less than 4 inches (102 mm). <u>Steel lintels shall be shop coated with a rust-inhibitive paint, except for lintels made of corrosion-resistant steel or steel treated with coatings to provide corrosion resistance. Construction of openings shall comply with either Section R703.7.3.1 or 703.7.3.2.</u>

R703.7.3.1 <u>The allowable span shall not exceed the values set forth in Table R703.7.3.1.</u>

R703.7.3.2 <u>The allowable span shall not exceed 18 feet 3 inches (5562 mm) and shall be constructed to comply with Figure R703.7.3.2 and the following:</u>

1. <u>Provide a minimum length of 18 inches (457 mm) of masonry veneer on each side of opening as shown in Figure R703.7.3.2.</u>

R703.7.3 continues

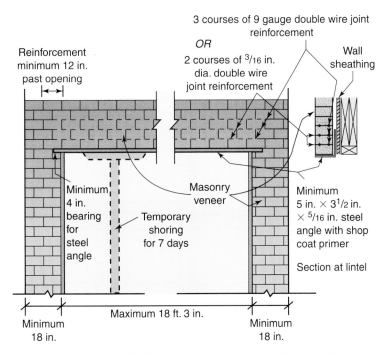

Masonry and steel lintel above overhead garage door opening

R703.7.3 continued

2. Provide a minimum 5-inch by 3½-inch by $^{5}/_{16}$-inch (127 mm by 89 mm by 7.9 mm) steel angle above the opening and shore for a minimum of 7 days after installation.

3. Provide double-wire joint reinforcement extending 12 inches (305 mm) beyond each side of opening. Lap splices of joint reinforcement a minimum of 12 inches (305 mm). Comply with one of the following:

    3.1. Double-wire joint reinforcement shall be $^{3}/_{16}$ inch (4.8 mm) diameter and shall be placed in the first two bed joints above the opening.

    3.2. Double-wire joint reinforcement shall be 9 gauge (0.144 inch or 3.66 mm diameter) and shall be placed in the first three bed joints above the opening.

CHANGE SIGNIFICANCE: The code requires noncombustible lintels to support masonry veneer that spans over openings such as doors and windows. Steel angles are commonly used as lintels to support masonry veneer above these openings. Typically, these steel lintels are shop coated with primer or other corrosion-resistant coating by the fabricator or supplier prior to delivery to the job site. The new text in Section R703.7.3 now specifically requires such corrosion resistance for steel lintels to inhibit the development of rust and protect the integrity of the masonry veneer. Corroded steel can expand significantly, resulting in stress cracking in the masonry veneer. Further corrosion may also decrease the load-bearing capacity of the lintel. Providing such protection after a steel lintel has been installed by painting the exposed portion of the lintel provides no protection to unexposed portions of the lintel.

The new Section R703.7.3.2 provides a cost-effective alternative to the existing steel lintel table for spanning large masonry veneer openings such as occur at overhead garage doors. These prescriptive provisions combine a steel angle with masonry veneer and reinforcing above to form the noncombustible lintel. The masonry veneer supported by the 5-inch × 3½-inch × $^{5}/_{16}$-inch steel angle acts as a beam when bonded together with horizontal joint reinforcement to span openings up to 18 feet 3 inches long. Shoring is required to support the steel lintel and veneer for a period of 7 days to allow the mortar to gain sufficient strength for the lintel to support the dead load of the masonry above.

CHANGE TYPE: Modification

CHANGE SUMMARY: The code now prescribes the minimum embedment and cover dimensions for metal wall ties in the mortar of masonry veneer.

2009 CODE: R703.7.4 Anchorage. Masonry veneer shall be anchored to the supporting wall with corrosion-resistant metal ties <u>embedded in mortar or grout and extending into the veneer a minimum of 1½ in. (38.1 mm), with not less than ⅝ inch (15.9 mm) mortar or grout cover to outside face.</u> Where veneer is anchored to wood backings by corrugated sheet metal ties, the distance separating the veneer from the sheathing material shall be a maximum of a nominal 1 inch (25 mm). Where the veneer is anchored to wood backings using metal strand wire ties, the distance separating the veneer from the sheathing material shall be a maximum of 4½ inches (114 mm). Where the veneer is anchored to cold-formed steel backings, adjustable metal strand wire ties shall be used. Where veneer is anchored to cold-formed steel backings, the distance separating the veneer from the sheathing material shall be a maximum of 4½ inches (114 mm).

CHANGE SIGNIFICANCE: The new text completes the necessary prescriptive requirements for anchorage of masonry veneer and provides consistency with ACI 530.1/ASCE 6/TMS 602 *Specification for Masonry Structures (MSJC Specification).* Whether constructed of solid or hollow masonry units, veneer requires wire or sheet metal ties to anchor it to the structure. Previous editions of the IRC specified the type, size, and spacing of the ties but lacked guidance on the embedment details. The code now prescribes a minimum embedment of 1½ inches into the mortar or grout with not less than ⅝-inch cover on the face side of the veneer.

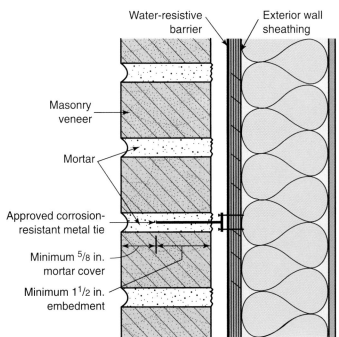

**Minimum embedment and mortar cover
dimensions for metal wall ties**

R703.11.1.1 and R703.11.2

Vinyl Siding

CHANGE TYPE: Addition

CHANGE SUMMARY: The 2009 IRC specifically requires vinyl soffit to be fastened in place in accordance with industry standards to ensure adequate wind resistance. New provisions address installations of vinyl siding over foam plastic sheathing based on design wind speed and wind exposure category.

2009 CODE: <u>**R703.11.1.1**   Soffit panels shall be individually fastened to a supporting component such as a nailing strip, fascia, or subfascia component or as specified by the manufacturer's instructions.</u>

<u>**R703.11.2 Foam Plastic Sheathing.**   Vinyl siding used with foam plastic sheathing shall be installed in accordance with Section R703.11.2.1, R703.11.2.2, or R703.11.2.3.</u>

> <u>**Exception:** Where the foam plastic sheathing is applied directly over wood structural panels, fiberboard, gypsum sheathing, or other approved backing capable of independently resisting the design wind pressure, the vinyl siding shall be installed in accordance with Section R703.11.1.</u>

<u>**R703.11.2.1 Basic Wind Speed Not Exceeding 90 mph and Exposure Category B.**   Where the basic wind speed does not ex-</u>

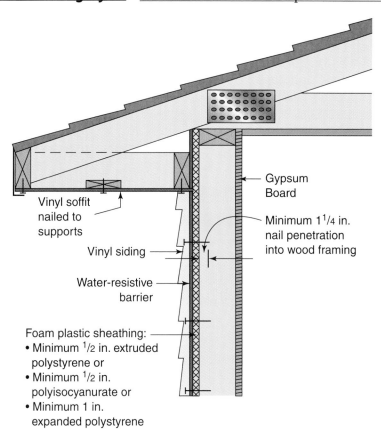

Vinyl siding installed over foam plastic sheathing, wind speed ≤90 mph, Exposure B

ceed 90 mph, the Exposure Category is B and gypsum wall board or equivalent is installed on the side of the wall opposite the foam plastic sheathing, the minimum siding fastener penetration into wood framing shall be 1¼ inches (32 mm) using minimum 0.120-inch-diameter nail (shank) with a minimum 0.313-inch-diameter head, 16 inches on center. The foam plastic sheathing shall be minimum ½-inch-thick (nominal) extruded polystyrene per ASTM C578, ½-inch-thick (nominal) polyisocyanurate per ASTM C1289, or 1-inch-thick (nominal) expanded polystyrene per ASTM C578.

R703.11.2.2 Basic Wind Speed Exceeding 90 mph or Exposure Categories C and D.   Where the basic wind speed exceeds 90 mph or the Exposure Category is C or D, or all conditions of R703.11.2.1 are not met, the adjusted design pressure rating for the assembly shall meet or exceed the loads listed in Tables R301.2(2) adjusted for height and exposure using R301.2(3). The design wind pressure rating of the vinyl siding for installation over solid sheathing as provided in the vinyl siding manufacturer's product specifications shall be adjusted for the following wall assembly conditions:

1. For wall assemblies with foam plastic sheathing on the exterior side and gypsum wall board or equivalent on the interior side of the wall, the vinyl siding's design wind pressure rating shall be multiplied by 0.39.
2. For wall assemblies with foam plastic sheathing on the exterior side and no gypsum wall board or equivalent on the interior side of wall, the vinyl siding's design wind pressure rating shall be multiplied by 0.27.

R703.11.2.3 Manufacturer Specification.   Where the vinyl siding manufacturer's product specifications provide an approved design wind pressure rating for installation over foam plastic sheathing, use of this design wind pressure rating shall be permitted and the siding shall be installed in accordance with the manufacturer's installation instructions.

CHANGE SIGNIFICANCE: Section R703.11.1 requires vinyl siding, soffit, and accessories to be installed in accordance with the manufacturer's installation instructions, requirements that have not changed in the 2009 IRC. To provide more guidance to the installer and code user, the IRC now specifies that vinyl soffit must be attached to suitable backing or nailing strips. The Vinyl Siding Institute's Installation Manual and most manufacturers specify such fastening for vinyl soffit panels to ensure adequate resistance to wind.

Section R703.11.2 has been added to improve wind-resistance performance for vinyl siding applied over foam plastic sheathing, a common installation for meeting energy-efficiency requirements. The code now offers prescriptive fastening requirements for areas with a basic wind speed not greater than 90 mph and a wind Exposure B condition. In addition to nail size and spacing requirements, these fastening specifications clarify that a minimum 1¼-inch siding nail penetration into the framing is required. The code provision allowing ¾ inch pen-

R703.11.1.1 and R703.11.2 continues

R703.11.1.1 and R703.11.2 continued

etration applies only when vinyl siding is installed over a solid substrate, such as wood structural panel sheathing, which is independently capable of resisting wind suction pressure.

For basic wind speeds greater than 90 mph or locations in Exposure Category C or D, the design wind pressure rating of the exterior wall covering assembly is determined by applying a prescribed adjustment factor to a base value in the vinyl siding manufacturer's product specifications. The higher adjustment factor is based on a wall assembly with gypsum board applied to the interior of the wall. The adjusted design pressure rating for the wall assembly must satisfy the component and cladding load requirements of Tables R301.2(2) and R301.2(3).

Both the prescriptive requirements and the adjustment factors are based on certified testing of various combinations of foam sheathing and vinyl siding products conducted at the National Association of Home Builders (NAHB) Research Center. The provisions also relied on testing serving as the basis for the wind pressure rating method for vinyl siding as explained in ASTM D 3679, Annex A.

CHANGE TYPE: Modification

CHANGE SUMMARY: Section R804 has been extensively revised and reorganized to clarify and update the prescriptive provisions for cold-formed steel light frame roof construction. The changes reflect the provisions in the new referenced standard AISI S230, *Standard for Cold-Formed Steel Framing—Prescriptive Method for One- and Two-Family Dwellings,* 2007 edition. Applicability of the prescriptive methods has expanded to include three-story buildings, an increase from the previous limitation of two stories.

2009 CODE: R804.1.1 Applicability Limits. The provisions of this section shall control the construction of cold-formed steel roof framing for buildings not greater than 60 feet (18 288 mm) perpendicular to the joist, rafter, or truss span, not greater than 40 feet (12 192 mm) in width parallel to the joist span or truss, ~~not greater than two~~ less than or equal to three stories above grade plane ~~in height~~ and with roof slopes not smaller than 3:12 (25 percent slope) or greater than 12:12 (100 percent slope). Cold-formed ~~S~~ steel roof framing constructed in accordance with the provisions of this section shall be limited to sites subjected to a maximum design wind speed of 110 miles per hour (49 m/s), Exposure ~~A,~~ B~~,~~ or C, and a maximum ground snow load of 70 pounds per square foot (3350 Pa).

R804.1.2 In-line Framing. Cold-formed ~~S~~ steel roof framing constructed in accordance with Section R804 shall be located ~~directly~~ in line with load-bearing studs in accordance with Figure R804.1.2 and the tolerances specified as follows.

1. ~~with a~~ The maximum tolerance shall be ~~of~~ ¾ inch (19.1 mm) between the centerline of the horizontal framing member and the centerline of the vertical framing member ~~between the center line of the stud and the roof joist/rafter.~~

2. Where the centerline of the horizontal framing member and bearing stiffener are located to one side of the center line of the vertical framing member, the maximum tolerance shall be

R804 continues

R804 continues

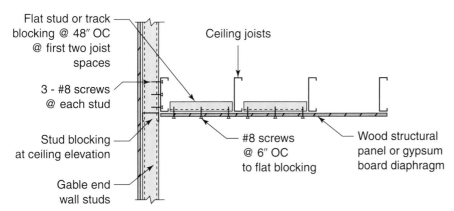

Flat stud or track blocking @ 48″ OC @ first two joist spaces

Ceiling joists

3 - #8 screws @ each stud

Stud blocking at ceiling elevation

#8 screws @ 6″ OC to flat blocking

Wood structural panel or gypsum board diaphragm

Gable end wall studs

Ceiling diaphragm to gable endwall detail

R804 continued

<u>$\frac{1}{8}$ inch (3 mm) between the web of the horizontal framing member and the edge of the vertical framing member.</u>

Because this code change deleted, added, and modified substantial portions of Section R804, the entire code change text is too extensive to be included here. Refer to the *2009 IRC Code Changes Resource Collection* for the complete text and history of code change RB209-07/08 related to Section R804.

CHANGE SIGNIFICANCE: The cold-formed steel roof framing provisions have been updated to correlate with the 2007 edition of AISI S230, *Standard for Cold-Formed Steel Framing—Prescriptive Method for One- and Two-Family Dwellings*, the new code-referenced standard of the American Iron and Steel Institute. The allowable number of stories has increased from two to three and now matches the height limitations of the prescriptive wood framing provisions.

All requirements concerning web holes and web hole adjustments are now consolidated in one location. The code user now has the choice to reinforce nonconforming holes, patch nonconforming holes, or design nonconforming holes in accordance with accepted engineering practice.

In place of *uncoated steel thickness*, the code now uses the current industry standardized thickness for structural members expressed as *base steel thickness* in mils (thousandths of an inch). *Reference Gage Number* is no longer used in referencing structural members and has been removed from the associated tables. Based on research, the in-line framing tolerance in Section R804.1.2 has been revised to account for the special case of the bearing stiffener located on the back-side of the joist.

Section R804.3.1, Ceiling Joists, has been modified to include the latest provisions from AISI S230-07 and to improve understanding of the application. Minimum ceiling joist size, ceiling joist bearing stiffeners, ceiling joist bottom flange bracing, ceiling joist top flange bracing, and ceiling joist splicing have been relocated into individual subsections to clarify the different requirements. In similar fashion, Section R804.3.2, Roof Rafters, places information for rafter size, rafter support brace, rafter splice, rafter to ceiling joist and ridge member connection, and rafter bottom flange bracing into separate subsections. New subsections on eave overhangs and rake overhangs have been added to the code to fill a current gap in information and to coordinate with AISI S230-07. In addition Figure R804.3.2.1.2 has been added to clarify this provision. The 2009 IRC contains new provisions for cold-formed steel hip framing to integrate the AISI provisions into the IRC and complete the prescriptive requirements in the roofing section.

The extensive changes to Section R804 also include new tables on roof rafter spans and for framing member and fastening requirements. New figures demonstrate the various connection details and provide information on screw types and a new detail for ceiling joist splices. Wind Exposure Category A has been deleted from the cold-formed steel provisions because it no longer exists in ASCE 7-05.

R806
Attic Ventilation

CHANGE TYPE: Modification

CHANGE SUMMARY: The attic ventilation requirements now permit methods and materials other than wire mesh for protecting openings against the entry of insects. The minimum opening dimension has been reduced from ⅛ to ¹⁄₁₆ inch. Vapor retarders are broken into three classes based on properties associated with the rate of restricting the passage of water. The provisions for unvented attic spaces have been rewritten for accuracy and clarification.

2009 CODE: **R806.1 Ventilation Required.**   Enclosed attics and enclosed rafter spaces formed where ceilings are applied directly to the underside of roof rafters shall have cross-ventilation for each separate space by ventilating openings protected against the entrance of rain or snow. Ventilation openings shall be provided with corrosion-resistant wire mesh, with 1/8 inch (3.2 mm) have a least dimension of ¹⁄₁₆ inch (1.6 mm) minimum to and ¼ inch (6.4 mm) maximum openings. Ventilation openings having a least dimension larger than ¼ inch (6.4 mm) shall be provided with corrosion-resistant wire cloth screening, hardware cloth, or similar material with openings having a least dimension of ¹⁄₁₆ inch (1.6 mm) minimum and ¼ inch (6.4 mm) maximum. Openings in roof framing members shall conform to the requirements of R802.7.

R806 continues

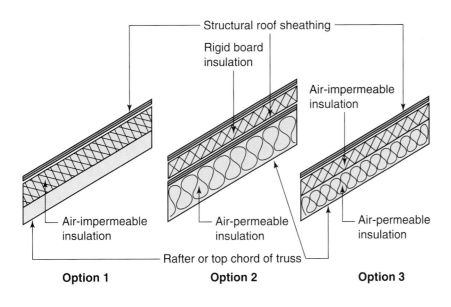

Structural roof sheathing

Rigid board insulation

Air-impermeable insulation

Air-impermeable insulation

Air-permeable insulation

Air-permeable insulation

Rafter or top chord of truss

Option 1   **Option 2**   **Option 3**

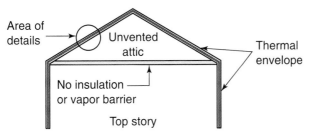

Area of details

Unvented attic

Thermal envelope

No insulation or vapor barrier

Top story

Unvented attic insulation

R806 continued

R806.2 Minimum Area. The total net free ventilating area shall not be less than $\frac{1}{150}$ of the area of the space ventilated except that reduction of the total area to $\frac{1}{300}$ is permitted, provided that at least 50 percent and not more than 80 percent of the required ventilating area is provided by ventilators located in the upper portion of the space to be ventilated at least 3 feet (914 mm) above the eave or cornice vents, with the balance of the required ventilation provided by eave or cornice vents. As an alternative, the net free cross-ventilation area may be reduced to $\frac{1}{300}$ when a ~~Class I or II~~ vapor barrier ~~having a transmission rate not exceeding 1 perm (5.7′ 10-11 kg/s × m2 × Pa)~~ is installed on the warm-in-winter side of the ceiling.

R806.3 Vent and Insulation Clearance. (No change to current text.)

R806.4 ~~Conditioned~~ Unvented Attic Assemblies. Unvented ~~conditioned~~ attic assemblies (spaces between the ceiling joists of the top story and the roof rafters) ~~are~~ shall be permitted ~~under~~ if all the following conditions are met:

1. The unvented attic space is completely contained within the building thermal envelope.

~~1.~~ 2. No interior vapor retarders are installed on the ceiling side (attic floor) of the unvented attic assembly.

~~2. An air-impermeable insulation is applied in direct contact to the underside/interior of the structural roof deck. "Air-impermeable" shall be defined by ASTM E 283.~~

   Exception: ~~In Zones 2B and 3B, insulation is not required to be air impermeable.~~

~~3. In the warm humid locations as defined in Section N1101.2.1:~~

~~3.1. For asphalt roofing shingles: A 1-perm (5.7′ 10-11 kg/s × m2 × Pa) or less vapor retarder (determined using Procedure B of ASTM E 96) is placed to the exterior of the structural roof deck; that is, just above the roof structural sheathing.~~

~~3.2.~~ 3. ~~For~~ Where wood shingles ~~and~~ or shakes are used, a minimum ~~continuous~~ $\frac{1}{4}$ inch (6 mm) vented air space separates the shingles ~~/~~or shakes and the roofing ~~felt placed over~~ underlayment above the structural sheathing.

~~4. In Zones 3 through 8 as defined in Section N1101.2, sufficient insulation is installed to maintain the monthly average temperature of the condensing surface above 45°F (7°C). The condensing surface is defined as either the structural roof deck or the interior surface of an air-impermeable insulation applied in direct contact with the underside/interior of the structural roof deck. "Air-impermeable" is quantitatively defined by ASTM E 283. For calculation purposes, an interior temperature of 68°F (20°C) is assumed. The exterior temperature is assumed to be the monthly average outside temperature.~~

4. In Climate Zones 5, 6, 7, and 8, any air-impermeable insulation shall be a vapor retarder, or shall have a vapor retarder coating or covering in direct contact with the underside of the insulation.

5. Either Items 5.1, 5.2, or 5.3 shall be met, depending on the air permeability of the insulation directly under the structural roof sheathing.

   5.1. Air-impermeable insulation only. Insulation shall be applied in direct contact to the underside of the structural roof sheathing.

   5.2. Air-permeable insulation only. In addition to the air-permeable installed directly below the structural sheathing, rigid board or sheet insulation shall be installed directly above the structural roof sheathing, as specified in Table R806.4 for condensation control.

   5.3. Air-impermeable and air-permeable insulation. The air-impermeable insulation shall be applied in direct contact to the underside of the structural roof sheathing as specified in Table R806.4 for condensation control. The air-permeable insulation shall be installed directly under the air-impermeable insulation.

SECTION R202 DEFINITIONS

Air-Impermeable Insulation.   An insulation having an air permanence equal to or less than 0.02 L/s-m^2 at 75 Pa pressure differential tested according to ASTM E 2178 or E 283.

~~**Vapor Retarder.**~~   ~~A vapor resistant material, membrane or covering such as foil, plastic sheeting, or insulation facing having a permeance rating of 1 perm (5.7 X 10-11 kg/Pa . E s .. Em2) or less when tested in accordance with the dessicant method using Procedure A of ASTM E 96. Vapor retarders limit the amount of moisture vapor that passes through a material or wall assembly.~~

Vapor Retarder Class.   A measure of the ability of a material or assembly to limit the amount of moisture that passes through that mate-

R806 continues

TABLE R806.4   **Insulation for Condensation Control**

Climate Zone	Minimum Rigid Board or Air-Impermeable Insulation *R*-Value[a]
2B and 3B tile roof only	0 (none required)
1, 2A, 2B, 3A, 3B, 3C	R-5
4C	R-10
4A, 4B	R-15
5	R-20
6	R-25
7	R-30
8	R-35

a. Contributes to but does not supersede Chapter 11 energy requirements.

R806 continued
rial or assembly. Vapor retarder class shall be defined using the desiccant method with Procedure A of ASTM E 96 as follows:

Class I: 0.1 perm or less
Class II: 0.1 < perm ≤ 1.0 perm
Class III: 1.0 < perm ≤ 10 perm

CHANGE SIGNIFICANCE: Roof louvers, ridge vents, soffit vents, and other openings provide required ventilation to attics. Previously, the code required metal wire mesh on these openings to prevent insects from entering the ventilated area. The change recognizes that modern manufacturing techniques produce products with punched, slotted, or hidden ventilation openings that do not require traditional insect screening. The opening dimensions are no longer limited to ¼ inch in both directions and now permit long narrow slots for ventilation purposes. The least dimension of the opening has been reduced from ⅛ to ¹⁄₁₆ inch to reflect current manufacturing practices. The reference to Section R802.7 reminds the code user that bored holes and notches in wood framing members are limited in size to preserve the structural integrity of the member. In addition to rafters, beams, and ceiling joists, the limitations also apply to required blocking between rafters at their bearing support, which may need to be bored for attic ventilation purposes.

The code now defines three classes of vapor retarders with varying degrees of moisture permeability. For the reduction of the net free cross-ventilation area in Section R806.2, a Class I or Class II vapor retarder must be used on the ceiling, permitting a higher permeability rate and greater flexibility in design than in the 2006 IRC. This change recognizes that many common materials function to various degrees to slow the passage of moisture. In many situations, common materials such as the kraft facing on fiberglass batts or latex paint may serve to retard moisture sufficiently and polyethylene sheeting, a common material with a lower perm rating, may not be required.

In unvented attics, the thermal envelope (insulation and air barrier boundary) is above the attic space rather than on top of the ceiling. Unvented attics eliminate the extreme temperatures of the attic, thereby providing a more favorable environment for HVAC equipment, ducts, and piping and offering a desirable design option for energy-efficient construction. The rewrite of Section R806.4 clarifies the requirements for unvented attics and modifies some provisions based on analysis and field experience. Unvented attics do not necessarily require a direct conditioned air source, and the word *conditioned* has been removed to correct a common misunderstanding of this section. The code now defines the performance characteristics of air impermeable insulation as required for unvented attics. All conditions of Items 1–5 must be satisfied for compliance with the provisions of Section R806.4. In Items 4 and 5, the code provisions have been simplified for better understanding. For example, calculations to determine the amount of insulation required to maintain the monthly average temperature of the condensing surface above 45°F are no longer necessary. Instead, Table R806.4 prescribes the minimum *R*-value of rigid board or air-impermeable insulation required to control condensation in unvented attics based on Climate Zone.

R807.1
Attic Access

CHANGE TYPE: Clarification

CHANGE SUMMARY: Section R807.1 now prescribes the methods to measure the height of attics requiring access and the height above the attic access opening.

2009 CODE: R807.1 Attic Access.   Buildings with combustible ceiling or roof construction shall have an attic access opening to attic areas that exceed 30 square feet (2.8 m^2) and have a vertical height of 30 inches (762 mm) or ~~more~~ greater. The vertical height shall be measured from the top of the ceiling framing members to the underside of the roof framing members.

The rough-framed opening shall not be less than 22 inches by 30 inches (559 mm by 762 mm) and shall be located in a hallway or other readily accessible location. When located in a wall, the opening shall be a minimum of 22 inches wide by 30 inches high. When the access is located in a ceiling, A ~~30-inch (762 mm)~~ minimum unobstructed headroom in the attic space shall be ~~provided~~ 30 inches (762 mm) at some point above the access ~~opening~~ measured vertically from the bottom of ceiling framing members. See Section M1305.1.3 for access requirements where mechanical equipment is located in attics.

CHANGE SIGNIFICANCE: The intent of this change is to resolve some confusion regarding the methods for measuring heights of attics and the required height above an attic access, and to promote uniform application of the provisions. The new text clarifies that measurements are taken from the framing members and not from the insulation. In determining attic height, the measurement is taken from the top of the ceiling joist or truss bottom chord to the bottom of the rafter or truss top chord. Conversely, the minimum clearance height above the attic access opening is measured from the bottom of the ceiling joist or truss bottom chord. The precise methods for measuring heights may be particularly helpful in low-slope roof applications. The other change to this section clarifies that access openings through a wall require a minimum height of 30 inches.

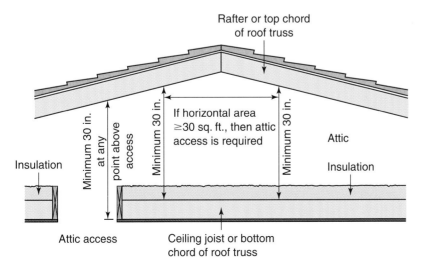

Measuring attic height and attic access headroom

R905.2

Asphalt Shingles

CHANGE TYPE: Modification

CHANGE SUMMARY: The changes to the asphalt shingle provisions clarify the attachment and wind-resistance requirements and correlate to the applicable ASTM standards. New tables provide asphalt shingle classifications based on design wind speed and whether the shingles are sealed in accordance with ASTM D 7158 or unsealed in accordance with ASTM D 3161. The valley lining provisions have been revised to reference the appropriate standard for the use of self-adhering polymer modified bitumen underlayment in a closed valley application. The code now prescribes the minimum dimensions for step-flashings. Editorial changes improve the mandatory language.

2009 CODE: R905.2.4 Asphalt Shingles. Asphalt shingles shall ~~have self-seal strips or be interlocking, and~~ comply with ASTM D 225 or D 3462.

R905.2.4.1 Wind Resistance of Asphalt Shingles. ~~Asphalt shingles shall be installed in accordance with Section R905.2.6. Shingles classified using ASTM D 3161 are acceptable for use in wind zones less than 110 mph (49 m/s). Shingles classified using ASTM D 3161, Class F, are acceptable for use in all cases where special fastening is required.~~ Asphalt shingles shall be tested in accordance with ASTM D 7158. Asphalt shingles shall meet the classification requirements of Table R905.2.4.1(1) for the appropriate maximum basic wind speed. Asphalt shingle packaging shall bear a label to indicate compliance with ASTM D7158 and the required classification in Table R905.2.4.1(1).

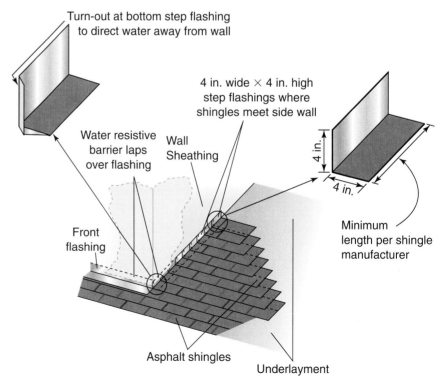

Turn-out at bottom step flashing to direct water away from wall

4 in. wide × 4 in. high step flashings where shingles meet side wall

4 in.

4 in.

Water resistive barrier laps over flashing

Wall Sheathing

Minimum length per shingle manufacturer

Front flashing

Asphalt shingles

Underlayment

Sidewall flashing for asphalt shingles

Exception: Asphalt shingles not included in the scope of ASTM D 7158 shall be tested and labeled to indicate compliance with ASTM D 3161 and the required classification in Table R905.2.4.1(2).

R905.2.6 Attachment. Asphalt shingles shall have the minimum number of fasteners required by the manufacturer. ~~For normal application, asphalt shingles shall be secured to the roof with,~~ but not less than four fasteners per strip shingle or two fasteners per individual shingle. Where the roof slope exceeds ~~20~~ 21 units vertical in 12 units horizontal (21:12, ~~167~~ 175 percent slope), ~~special methods of fastening are required~~ shingles shall be installed as required by the manufacturer. ~~For roofs located where the basic wind speed per Figure R301.2(4) is 110 mph (49 m/s) or higher, special methods of fastening are required. Special fastening methods shall be tested in accordance with ASTM D 3161, Class F. Asphalt shingle wrappers shall bear a label indicating compliance with ASTM D 3161, Class F.~~

R905.2.8.2 Valleys. Valley linings shall be installed in accordance with the manufacturer's installation instructions before applying shingles. Valley linings of the following types shall be permitted:

TABLE R905.2.4.1(1)   Classification of Asphalt Shingles per ASTM D 7158

Maximum Basic Wind Speed from Figure R301.2(4)	Classification Requirement
85	D, G, or H
90	D, G, or H
100	G or H
110	G or H
120	G or H
130	H
140	H
150	H

TABLE R905.2.4.1(2)   Classification of Asphalt Shingles per ASTM D 3161

Maximum Basic Wind Speed From Figure R301.2(4)	Classification Requirement
85	A, D, or F
90	A, D, or F
100	A, D, or F
110	F
120	F
130	F
140	F
150	F

R905.2 continues

R905.2 continued

1. For open valley~~s~~ (valley lining exposed) lined with metal, the valley lining shall be at least 24 inches (610 mm) wide and of any of the corrosion-resistant metals in Table R905.2.8.2.

2. For open valleys, valley lining of two plies of mineral surfaced roll roofing, complying with ASTM D 3909 or ASTM D 6380 Class M, shall be permitted. The bottom layer shall be 18 inches (457 mm) and the top layer a minimum of 36 inches (914 mm) wide.

3. For closed valleys (valley covered with shingles), valley lining of one ply of smooth roll roofing complying with ASTM D 6380 ~~Class S Type III, Class M Type II, or ASTM D 3909~~ and at least 36 inches wide (914 mm) or valley lining as described in Items 1 or 2 above shall be permitted. ~~Specialty~~ <u>Self-adhering polymer modified bitumen</u> underlayment complying with ASTM D 1970 ~~may be used~~ <u>shall be permitted</u> in lieu of the lining material.

R905.2.8.3 Sidewall Flashing. Flashing against a vertical sidewall shall be by the step-flashing method. <u>The flashing shall be a minimum of 4 inches (102 mm) high and 4 inches (102 mm) wide. At the end of the vertical sidewall, the step flashing shall be turned out in such a manner so as to direct water away from the wall and onto the roof and/or gutter.</u>

CHANGE SIGNIFICANCE: Revisions to the asphalt shingle provisions define a clear path for determining compliance and bring requirements in line with the appropriate referenced standards. The change to the charging statement in Section R905.2.4 removes unnecessary language related to self-sealing or interlocking of shingles. Such terms are undefined and are unnecessary with the introduction of requirements for testing to ASTM D 7158 or ASTM D 3161 in the wind-resistance provisions.

Section R905.2.4.1 now provides clear scoping for the applicable test standards in demonstrating resistance to wind forces for asphalt shingles. The new tables assist in the proper selection of asphalt shingle classification based upon the basic wind speed and the applicable standard. ASTM D 7158 *Standard Test Method for Wind Resistance of Sealed Asphalt Shingles* has been added as an approved referenced standard in the IRC. Self-sealing shingles must meet the test requirements of ASTM D7158 and be labeled in accordance with Table R905.2.4.1(1). ASTM D 3161 *Standard Test Method for Wind-Resistance of Asphalt Shingles* applies only to shingles that are not sealed to provide resistance to wind uplift. Therefore, unsealed asphalt shingles must meet the classification requirements in Table R905.2.4.1(2).

The attachment section retains the requirements for the minimum number of fasteners for asphalt shingles provided that number also satisfies the manufacturer's requirements. The methods for fastening shingles in high slope applications (slopes exceeding 21:12) must comply with the manufacturer's installation instructions. The provisions containing subjective language related to special methods of fastening and wind speed have been deleted. Products must conform to the applicable standards of Section R905.2.4 and meet the classification requirements of the new tables based on wind speed. Installation and attachment, including any special provisions for high wind areas, must meet the manufacturer's requirements.

Item 3 of Section R905.2.8.2 has been revised to clarify the closed valley lining provisions. The previous referenced standard, ASTM D3909, more appropriately applies to materials used in open valley conditions and has been deleted from Item 3. The added language clarifies that ASTM D1970 applies to self-adhering modified bitumen underlayment and is now consistent with the underlayment requirements in Section R905.2.3.

This code change also introduces minimum dimensions for step-flashings at the junction of asphalt shingles and sidewalls, reflecting industry recommendations. Depending on the type of siding, the minimum clearance above the roof surface may be as much as 2 inches and the minimum 4-inch flashing height allows for adequate lap over the flashing by the water-resistive barrier and the siding. This section also addresses the lower end termination of the sidewall flashing to divert away from the wall and onto the roofing material.

R905.8.6

Wood Shake Installation

CHANGE TYPE: Modification

CHANGE SUMMARY: To improve longevity, the minimum spacing between wood shakes has increased to $^3/_8$ inch and is now consistent with the Cedar Shake and Shingle Bureau's application instructions.

2009 CODE: R905.8.6 Application. Wood shakes shall be installed according to this chapter and the manufacturer's installation instructions. Wood shakes shall be laid with a side lap not less than 1½ inches (38 mm) between joints in adjacent courses. Spacing between shakes in the same course shall be ~~1/8~~ $^3/_8$ inch to $^5/_8$ inch (~~3~~ 9.5 mm to ~~16~~ 15.9 mm) for shakes and taper sawn shakes of naturally durable wood and shall be ~~¼~~ $^3/_8$ inch to ~~3/8~~ $^5/_8$ inch (~~6~~ 9.5 mm to ~~10~~ 15.9 mm) for preservative-treated taper sawn shakes. Weather exposure for wood shakes shall not exceed those set forth in Table R905.8.6. Fasteners for wood shakes shall be corrosion resistant, with a minimum penetration of ½ inch (12.7 mm) into the sheathing. For sheathing less than ½ inch (12.7 mm) in thickness, the fasteners shall extend through the sheathing. Wood shakes shall be attached to the roof with two fasteners per shake, positioned no more than 1 inch (25 mm) from each edge and no more than 2 inches (51 mm) above the exposure line.

CHANGE SIGNIFICANCE: The Cedar Shake and Shingle Bureau's application instructions require the space between adjacent wood shakes, called the *keyway*, to be not less than $^3/_8$ inch. Previously, the code permitted keyway widths as small as $^1/_8$ inch. Field observation indicates that such narrow spacing traps leaves or other organic debris and may cause premature aging of wood shakes. This change brings the code language into agreement with manufacturer's recommendations and intends to increase the life of a wood shake roof.

CHANGE TYPE: Modification

CHANGE SUMMARY: The IRC adds minimum thickness, parging, and lining requirements to the masonry fireplace smoke chamber provisions and references the applicable standards. The revised definition of masonry chimney provides consistency with language in the masonry fireplace and smoke chamber sections. Clay flue liners for masonry chimneys require a non-water-soluble refractory mortar in accordance with ASTM C 1283 and ASTM C 199.

2009 CODE: R1001.8 Smoke Chamber.   Smoke chamber walls shall be constructed of solid masonry units, hollow masonry units grouted solid, stone or concrete. <u>The total minimum thickness of front, back, and side walls shall be 8 inches (203 mm) of solid masonry. The inside surface shall be parged smooth with refractory mortar conforming to ASTM C 199.</u> ~~Corbelling of masonry units shall not leave unit cores exposed to the inside of the smoke chamber.~~ When a lining of firebrick at least 2 inches (51 mm) thick, or a lining of vitrified clay at least ⅝ inch (16 mm) thick, is provided, the total minimum thickness of front, back and side walls shall be 6 inches (152 mm) of solid masonry, including the lining. Firebrick shall conform to ASTM ~~C 27 or~~ C 1261 and shall be laid with medium duty refractory mortar conforming to ASTM C 199. <u>Vitrified clay linings shall conform to ASTM C 315.</u> ~~Where no lining is provided, the total minimum thickness of~~

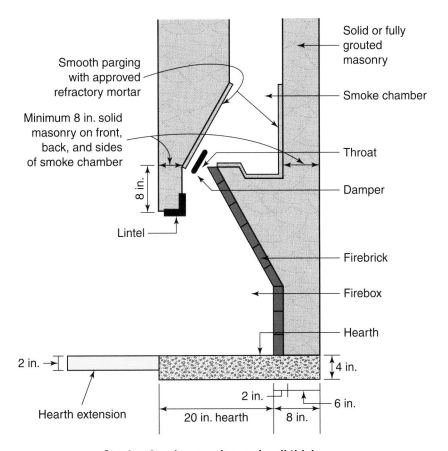

Smoke chamber parging and wall thickness

Labels in figure:
- Solid or fully grouted masonry
- Smoke chamber
- Smooth parging with approved refractory mortar
- Throat
- Minimum 8 in. solid masonry on front, back, and sides of smoke chamber
- 8 in.
- Damper
- Lintel
- Firebrick
- Firebox
- Hearth
- 2 in.
- Hearth extension
- 2 in.
- 4 in.
- 6 in.
- 20 in. hearth
- 8 in.

R1001 and R1003

Masonry Fireplaces and Chimneys

~~front, back and side walls shall be 8 inches (203 mm) of solid masonry. When the inside surface of the smoke chamber is formed by corbeled masonry, the inside surface shall be parged smooth.~~

R1003.1 Definition.   A masonry chimney is a chimney constructed of ~~concrete or masonry~~ <u>solid masonry units, hollow masonry units grouted solid, stone or concrete</u>, hereinafter referred to as masonry. Masonry chimneys shall be constructed, anchored, supported, and reinforced as required in this chapter.

R1003.12 Clay Flue Lining (Installation).   Clay flue liners shall be installed in accordance with ASTM C 1283 and extend from a point not less than 8 inches (203 mm) below the lowest inlet or, in the case of fireplaces, from the top of the smoke chamber to a point above the enclosing walls. The lining shall be carried up vertically, with a maximum slope no greater than 30 degrees (0.52 rad) from the vertical.

Clay flue liners shall be laid in medium-duty <u>water insoluble</u> refractory mortar conforming to ASTM C 199 with tight mortar joints left smooth on the inside and installed to maintain an air space or insulation not to exceed the thickness of the flue liner separating the flue liners from the interior face of the chimney masonry walls. Flue liners shall be supported on all sides. Only enough mortar shall be placed to make the joint and hold the liners in position.

CHANGE SIGNIFICANCE: Masonry fireplace smoke chambers now specifically require protection with parging or clay flue liners able to withstand temperatures of 1800 degrees Fahrenheit as required of all the other fireplace and chimney lining materials. The new text in this section also references the appropriate ASTM standards and intends to preserve the integrity of masonry fireplaces. The reference to protecting the cores of corbeled masonry units has been removed because the required parging with refractory mortar ensures that these units will be covered.

The revised definition for masonry chimney clarifies that approved materials include both solid and hollow masonry units and stone in addition to concrete. Hollow masonry units must be grouted solid. These materials are consistent with those approved for masonry fireplaces in Section R1001.5, Firebox Walls, and R1001.8, Smoke Chamber.

Revisions to ASTM C 1283 *Standard Practice for Installing Clay Flue Lining* requires water insoluble refractory mortar for setting clay flue linings. The corresponding requirement in Section R1003.12 alerts code users that because flue linings are exposed to the weather, the refractory mortar must be water insoluble, conforming to ASTM C 199 *Standard Test Method for Pier Test for Refractory Mortars.*

PART 4

Energy Conservation
Chapter 11

Part 4 addresses the issues of residential building energy efficiency and energy conservation. Although IRC energy provisions are specifically focused on one- and two-family dwellings and townhouses, the provisions contained within the International Energy Conservation Code (IECC) can also be used to design the building envelope and the other systems related to energy efficiency for any building regulated under the IRC.

Section N1101 establishes climate zones for geographical locations as the basis for determining thermal envelope requirements for conserving energy under Chapter 11 of the IRC. A permanent energy certificate listing the values of installed insulation, fenestration, and equipment is required by Section N1101.9. The various elements of the building thermal envelope are covered in Section N1102 and include specific insulation, fenestration, air leakage, and moisture control requirements for improving energy efficiency. Section N1103 is concerned primarily with mechanical system controls, the insulation and sealing of ductwork in unconditioned spaces, and the insulation of mechanical piping systems. Energy-efficient lighting is covered in Section N1104. ■

N1101.2.1 AND TABLE N1101.2

Climate Zones, Moisture Regimes, and Warm-Humid Designations

N1101.9

Permanent Energy Certificate

TABLES N1102.1 AND N1102.1.2

Insulation and Fenestration Requirements by Component

N1102.2

Ceiling and Access Hatch Insulation Requirements

TABLE N1102.2.5

Steel Framed Wall Insulation

N1102.4.1

Sealing of the Building Thermal Envelope

N1102.4.5

Recessed Lighting

N1103.7

Snow Melting System Controls

N1103.8

Swimming Pools

N1104

Lighting Systems

N1101.2.1 and Table N1101.2

Climate Zones, Moisture Regimes, and Warm-Humid Designations

CHANGE TYPE: Clarification

CHANGE SUMMARY: Tables N1101.2 and N1101.2.1 have been combined into one table to include climate zones, moisture regimes, and warm-humid designations for every county.

2009 CODE: N1101.2.1 Warm Humid Counties. Warm humid counties are ~~listed~~ identified in Table N1101.2.~~1~~ by an asterisk.

~~**TABLE N1101.2**~~
~~Climate Zones by States and Counties~~

~~**TABLE N1101.2.1**~~
~~Warm Humid Counties~~

CHANGE SIGNIFICANCE: This change places climate zones, moisture regimes, and warm-humid designations in one table and provides designations for every county to add clarity and improve usability. Previously, counties with the predominant climate zone or warm-humid designation for a state were not individually listed, but noted as the general rule for the state. The code user had to scan the list of counties and if the county was not listed, refer to the default value for the state. Further, the existing format required such lookups in two separate tables to determine both zone/moisture regime and warm-humid status for southeastern locations. Warm-humid counties are now indicated with an asterisk (*) in Table N1101.2.

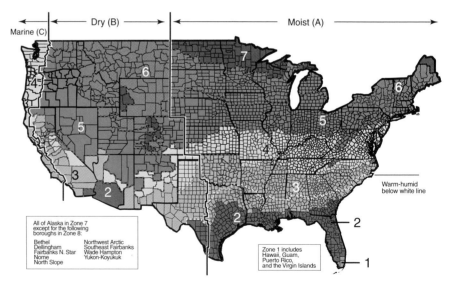

Cilmate zone map

TABLE N1101.2 (Excerpt)   Climate Zones, Moisture Regimes, and Warm Humid Designations by State, County, and Territory

Key:
A = Moist, B = Dry, C = Marine
Absence of moisture designation indicates moisture regime is irrelevant. Asterisk (*) indicates a warm-humid location.

ALABAMA	ALASKA	ARKANSAS	CALIFORNIA
3A Autauga*	7 Aleutians East	3A Arkansas	3C Alameda
2A Baldwin*	7 Aleutians West	3A Ashley	6B Alpine
3A Barbour*	7 Anchorage	4A Baxter	4B Amador
3A Bibb	8 Bethel	4A Benton	3B Butte
3A Blount	7 Bristol Bay	4A Boone	4B Calaveras
3A Bullock*	7 Denali	3A Bradley	3B Colusa
3A Butler*	8 Dillingham	3A Calhoun	3B Contra Costa
3A Calhoun	8 Fairbanks North Star	4A Carroll	4C Del Norte
3A Chambers	7 Haines	3A Chicot	4B El Dorado
3A Cherokee	7 Juneau	3A Clark	3B Fresno
3A Chilton	7 Kenai Peninsula	3A Clay	3B Glenn
3A Choctaw*	7 Ketchikan Gateway	3A Cleburne	4C Humboldt
3A Clarke*	7 Kodiak Island	3A Cleveland	2B Imperial
3A Clay	7 Lake and Peninsula	3A Columbia*	4B Inyo
3A Cleburne	7 Matanuska-Susitna	3A Conway	3B Kern
3A Coffee*	8 Nome	3A Craighead	3B Kings
3A Colbert	8 North Slope	3A Crawford	4B Lake
3A Conecuh*	8 Northwest Arctic	3A Crittenden	5B Lassen
3A Coosa	7 Prince of Wales-Outer Ketchikan	3A Cross	3B Los Angeles
3A Covington*	7 Sitka	3A Dallas	3B Madera
3A Crenshaw*	7 Skagway-Hoonah-Angoon	3A Desha	3C Marin
3A Cullman	8 Southeast Fairbanks	3A Drew	4B Mariposa
3A Dale*	7 Valdez-Cordova	3A Faulkner	3C Mendocino
3A Dallas*	8 Wade Hampton	3A Franklin	3B Merced
3A DeKalb	7 Wrangell-Petersburg	4A Fulton	5B Modoc
3A Elmore*	7 Yakutat	3A Garland	6B Mono
3A Escambia*	8 Yukon-Koyukuk	3A Grant	3C Monterey
3A Etowah	**ARIZONA**	3A Greene	3C Napa
3A Fayette	5B Apache	3A Hempstead*	5B Nevada
3A Franklin	3B Cochise	3A Hot Spring	3B Orange

N1101.2.1 and Table N1101.2 continues

N1101.2.1 and Table N1101.2 continued

TABLE N1101.2 (Excerpt)   **Climate Zones, Moisture Regimes, and Warm Humid Designations by State, County, and Territory**

Key:

A = Moist, B = Dry, C = Marine

Absence of moisture designation indicates moisture regime is irrelevant. Asterisk (*) indicates a warm-humid location.

ALABAMA	ARIZONA	ARKANSAS	CALIFORNIA
3A Geneva*	5B Coconino	3A Howard	3B Placer
3A Greene	4B Gila	3A Independence	5B Plumas
3A Hale	3B Graham	4A Izard	3B Riverside
3A Henry*	3B Greenlee	3A Jackson	3B Sacramento
3A Houston*	2B La Paz	3A Jefferson	3C San Benito
3A Jackson	2B Maricopa	3A Johnson	3B San Bernardino
3A Jefferson	3B Mohave	3A Lafayette*	3B San Diego
3A Lamar	5B Navajo	3A Lawrence	3C San Francisco
3A Lauderdale	2B Pima	3A Lee	3B San Joaquin
3A Lawrence	2B Pinal	3A Lincoln	3C San Luis Obispo

CHANGE TYPE: Modification

CHANGE SUMMARY: This change clarifies that the permanent energy certificate cannot cover the service directory or other required information on the electrical panel. When applied to gas-fired unvented heaters, electric furnaces, and baseboard heaters, energy-efficiency ratings are considered misleading. Installation of such appliances must be specifically noted on the energy certificate without reference to an efficiency designation.

N1101.9

Permanent Energy Certificate

Energy Efficiency Certificate		
Insulation Rating		***R*-Value**
Ceiling/roof		
Walls	Frame	
	Mass	
	Basement	
	Crawl space	
Floors	Over unconditioned space	
	Slab edge	
Ducts	Outside conditioned spaces	
Glass and Door Rating	**NFRC U-Factor**	**NFRC SHGC**
Window		
Opaque door		
Skylight		
Heating and Cooling Equipment	**Type**	**Efficiency**
Heating system		AFUE
Cooling system		SEER
Water heater		EF

Indicate if the following have been installed (an efficiency shall not be listed):

☐ electric furnace   ☐ gas-fired unvented room heater   ☐ baseboard electric heater

Designer _____

Builder _____

Date _____

Permanent energy certificate

N1101.9 continues

N1101.9 continued

2009 CODE: N1101.9 Certificate. A permanent certificate shall be posted on or in the electrical distribution panel. The certificate shall not cover or obstruct the visibility of the circuit directory label, service disconnect label or other required labels. The certificate shall be completed by the builder or registered design professional. The certificate shall list the predominant *R*-values of insulation installed in or on ceiling/roof, walls, foundation (slab, basement wall, crawlspace wall and/or floor) and ducts outside conditioned spaces; *U*-factors for fenestration; and the solar heat gain coefficient (SHGC) of fenestration. Where there is more than one value for each component, the certificate shall list the value covering the largest area. The certificate shall list the types and efficiencyies of heating, cooling and service water heating equipment. Where a gas-fired unvented room heater, electric furnace, and/or baseboard electric heater is installed in the residence, the certificate shall list gas-fired unvented room heater, electric furnace, or baseboard electric heater as appropriate. An efficiency shall not be listed for gas-fired unvented room heaters, electric furnaces, or electric baseboard heaters.

CHANGE SIGNIFICANCE: The electrical provisions of the IRC require a circuit directory and the manufacturer's identification information on the electrical service panel. The energy certificate does not govern such safety information that may be provided or required on the panel. The new language in Section N1101.9 clarifies that the certificate cannot cover or obscure information or labels associated with the electrical equipment.

Because energy-efficiency ratings for electric furnaces, baseboard heaters, and unvented gas-fired heaters may be misleading, the 2009 IRC requires such appliances to be individually listed on the certificate without an efficiency designation. Unvented gas-fired heaters typically serve a small area of the dwelling, vent the moisture they produce into the residence, and are not designed or intended to serve as the primary comfort heating source. Though they may have a higher rating than a high-efficiency central furnace, from an energy conservation standpoint the furnace is typically more efficient in heating the dwelling. Similarly, electric furnaces and baseboard heaters, while efficient at turning electricity into heat, may not provide the lowest energy consumption when compared with other methods of comfort heating.

CHANGE TYPE: Modification

CHANGE SUMMARY: For Climate Zones 5, 6, and Marine 4, the minimum wood frame wall *R*-value has been increased from R-19 to R-20, and the corresponding *U*-factors have been lowered to 0.057. Changes to the mass wall provisions simplify the format and align the IRC with correlating provisions in the *International Energy Conservation Code* (IECC). *U*-factors for fenestration have been lowered in Climate Zones 2, 3, and 4 to improve energy efficiency. An exception allows a higher *U*-factor for impact-resistant glazing required in windborne debris regions.

Tables N1102.1 and N1102.1.2

Insulation and Fenestration Requirements by Component

2009 CODE:

N1102.1.2 *U*-Factor Alternative. An assembly with a *U*-factor equal to or less than that specified in Table N1102.1.2 shall be permitted as an alternative to the *R*-value in Table N1102.1.

> **Exception:** ~~For mass walls not meeting the criterion for insulation location in Section N1102.2.3, the *U*-factor shall be permitted to be:~~
> 1. ~~*U*-factor of 0.17 in Climate Zone 1~~
> 2. ~~*U*-factor of 0.14 in Climate Zone 2~~
> 3. ~~*U*-factor of 0.12 in Climate Zone 3~~
> 4. ~~*U*-factor of 0.10 in Climate Zone 4 except Marine~~
> 5. ~~*U*-factor of 0.082 in Climate Zone 5 and Marine 4~~

N1102.2.4 Mass Walls. Mass walls, for the purposes of this chapter, shall be considered <u>above-grade</u> walls of concrete block, concrete, insulated concrete form (ICF), masonry cavity, brick (other than brick

Tables N1102.1 and N1102.1.2 continues

Climate zone	Minimum total insulation R-value		
3	5	5	8
4, except marine	5	5	10
5 and marine 4	13	13	17
6	15	15	19

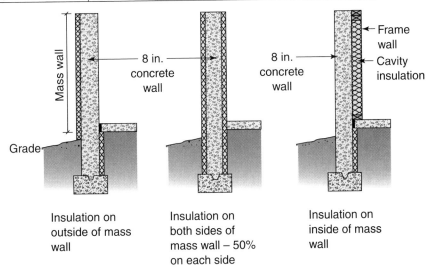

Insulation on outside of mass wall

Insulation on both sides of mass wall – 50% on each side

Insulation on inside of mass wall

Mass walls

TABLE N1102.1   Insulation and Fenestration Requirements by Component[a]

Climate Zone	Fenestration U-factor	Skylight[b] U-factor	Glazed Fenestra-tion SHGC	Ceiling R-value	Wood Frame Wall R-value	Mass Wall R-value[k]	Floor R-value	Basement[c] Wall R-value	Slab[d] R-value & Depth	Crawl Space[c] Wall R-value
1	1.20	0.75	0.40[j]	30	13	3/4̲	13	0	0	0
2	~~0.75~~ 0.65[i]	0.75	0.40[j]	30	13	4/6̲	13	0	0	0
3	~~0.65~~ 0.50[i]	0.65	0.40[e, j]	30	13	5/8̲	19	5/13̲[f]	0	5/13
4 except Marine	~~0.40~~ 0.35	0.60	NR	38	13	5/10̲	19	10/13	10, 2 ft	10/13
5 and Marine 4	0.35	0.60	NR	38	~~19~~ 20̲ or 13+5[h]	13/17̲	30[g]	10/13	10, 2 ft	10/13
6	0.35	0.60	NR	49	~~19~~ 20̲ or 13+5[h]	15/19̲	30[g]	10/13	10, 4 ft	10/13
7 and 8	0.35	0.60	NR	49	21	19/21̲	30[g]	10/13	10, 4 ft	10/13

a. [through] e. [no change to text]

f. Basement wall insulation is not required in warm-humid locations as defined by Figure N1101.2 and Table N1101.2.

~~f.~~ g̲. Or insulation sufficient to fill the framing cavity, R-19 minimum.

~~g.~~ h̲. "13+5" means R-13 cavity insulation plus R-5 insulated sheathing. If structural sheathing covers 25% or less of the exterior, R-5 sheathing is not required where structural sheathing is used. If structural sheathing covers more than 25% of exterior, structural sheathing shall be supplemented with insulated sheathing of at least R-2.

i. For impact-rated fenestration complying with Section R301.2.1.2, the maximum U-factor shall be 0.75 in zone 2 and 0.65 in zone 3.

j. For impact-resistant fenestration complying with Section R301.2.1.2, the maximum SHGC shall be 0.40.

k. The second R-value applies when more than half the insulation is on the interior.

TABLE N1102.1.2   Equivalent U-Factors[a]

Climate Zone	Fenestration U-factor	Skylight U-factor	Ceiling U-factor	Frame Wall U-factor	Mass Wall U-factor[b]	Floor U-factor	Basement Wall U-factor	Crawl Space Wall U-factor
1	1.20	0.75	0.035	0.082	0.197	0.064	0.360	0.477
2	~~0.75~~ 0.65̲	0.75	0.035	0.082	0.165	0.064	0.360	0.477
3	~~0.65~~ 0.50̲	0.65	0.035	0.082	0.141	0.047	0.091̲[c]	0.136
4 except Marine	~~0.40~~ 0.35̲	0.60	0.030	0.082	0.141	0.047	0.059	0.065
5 and Marine 4	0.35	0.60	0.030	~~0.060~~ 0.057̲	0.082	0.033	0.059	0.065
6	0.35	0.60	0.026	~~0.060~~ 0.057̲	0.060	0.033	0.059	0.065
7 and 8	0.35	0.60	0.026	0.057	0.057	0.033	0.059	0.065

a. Nonfenestration U-factors shall be obtained from measurement, calculation, or an approved source.

b. When more than half the insulation is on the interior, the mass wall U-factors shall be a maximum of 0.17 in Zone 1, 0.14 in Zone 2, 0.12 in Zone 3, 0.10 in Zone 4 except Marine, and the same as the frame wall U-factor in Marine Zone 4 and Zones 5 through 8.

c. Basement wall U-factor of 0.360 in warm-humid climates as defined by Figure N1101.2 and Table N1101.2.

Tables N1102.1 and N1102.1.2 continued

veneer), earth (adobe, compressed earth block, rammed earth) and solid timber/logs. ~~The provisions of Section N1102.1 for mass walls shall be applicable when at least 50 percent of the required insulation R-value is on the exterior of, or integral to, the wall. Walls that do not meet this criterion for insulation placement shall meet the wood frame wall insulation requirements of Section N1102.1.~~

Exception: ~~For walls that do not meet this criterion for insulation placement, the minimum added insulation *R*-value shall be permitted to be:~~

1. ~~*R*-value of 4 in Climate Zone 1~~

2. ~~*R*-value of 6 in Climate Zone 2~~

3. ~~*R*-value of 8 in Climate Zone 3~~

4. ~~*R*-value of 10 in Climate Zone 4 except Marine~~

5. ~~*R*-value of 13 in climate Zone 5 and Marine 4~~

CHANGE SIGNIFICANCE: The maximum *U*-factor for fenestration (windows, doors, and skylights) in Climate Zones 2 and 3 has been lowered to increase energy savings in these warmer regions. The new values are consistent with those for windows currently being installed in these areas due to the solar heat gain coefficient (SHGC) requirements. SHGC measures the effectiveness of a window or skylight to block heat gain from sunlight. The lower a window's SHGC, the less solar heat is transmitted. Low-E coatings are routinely used to achieve a low SHGC, which leads to a double pane window typically meeting the new *U*-factor values in Zones 2 and 3. The lower *U*-factors improve energy efficiency primarily in the heating season, which is particularly important in the northern areas of Zone 3, and this change is considered a cost-effective way to achieve significant energy savings. The new footnote *h* provides an exception to permit the current fenestration *U*-factors (0.75 for Zone 2 and 0.65 for Zone 3) in windborne debris regions when impact-resistant glazing is used to satisfy the opening protection requirements of Section R301.2.1.2. Because cooling loads are overwhelmingly predominant in Climate Zone 1, no benefit could be determined for lowering the *U*-factor in this zone, and the 0.120 value has been maintained. In Zone 4, where *U*-factor is most important to energy efficiency due to higher heating loads, the *U*-factor has been lowered to 0.35 to match the requirements of the International Energy Conservation Code (IECC).

In the colder Climate Zones 5 through 8 and Marine 4, the wood frame wall cavity *R*-value has been increased from R-19 to R-20 to increase energy savings during the heating season in these regions. This change reflects improvement and availability of R-20 insulation products that fit into 2 × 6 wall cavities without compression or decrease of the effective *R*-value. Because R-20 can be achieved with sprayed, blown, or batt products, it is not considered a *proprietary* value. The corresponding *U*-factor values of Table N1102.1.2 have been changed from 0.060 to 0.057 accordingly.

To take full advantage of the thermal mass properties of concrete, masonry, earth, and log walls in conserving energy, the insulation typically should be on the exterior side of or integral with the wall. The IRC requires higher *R*-values and lower equivalent *U*-values when more than 50 percent of the insulation occurs on the inside of the wall. Previously, the required values for walls meeting the interior insulation criterion were found in Sections N1102.1.2 and N1102.2.3. This change is one of formatting, simplifying the provisions by deleting text from the above-mentioned sections and placing the provisions in Tables N1102.1 and N1102.1.2. The IRC format is now consistent with that of the IECC.

N1102.2

Ceiling and Access Hatch Insulation Requirements

CHANGE TYPE: Clarification

CHANGE SUMMARY: This change clarifies that the thermal envelope requirements apply to hatches and doors that access unconditioned areas such as attics and crawl spaces. The provisions for reduced *R*-values in the ceiling insulation sections apply to only the prescriptive requirements of Table N1102.1 and do not apply to the *U*-factor or total UA alternatives.

2009 CODE: N1102.2 Specific insulation Requirements.
N1102.2.1 Ceilings with Attic Spaces. When Section N1102.1 would require R-38 in the ceiling, R-30 shall be deemed to satisfy the requirement for R-38 wherever the full height of uncompressed R-30 insulation extends over the wall top plate at the eaves. Similarly R-38 shall be deemed to satisfy the requirement for R-49 wherever the full height of uncompressed R-38 insulation extends over the wall top plate at the eaves. <u>This reduction shall not apply to the *U*-factor alternative approach in Section N1102.1.2 and the Total UA alternative in Section N1102.1.3.</u>

N1102.2.2 Ceilings without Attic Spaces. Where Section N1102.1 would require insulation levels above R-30 and the design of the roof/ceiling assembly does not allow sufficient space for the required insulation, the minimum required insulation for such roof/ceiling assemblies shall be R-30. This reduction of insulation from the requirements of Section N1102.1.<u>1</u> shall be limited to 500 square feet (46 m^2) of ceiling area. <u>This reduction shall not apply to the *U*-factor alternative approach in Section N1102.1.2 and the Total UA alternative in Section N1102.1.3.</u>

<u>N1102.2.3 Access Hatches and Doors.</u> <u>Access doors from conditioned spaces to unconditioned spaces (e.g., attics and crawl spaces) shall be weatherstripped and insulated to a level equivalent to the insulation on the surrounding surfaces. Access shall be provided to all equipment which prevents damaging or compressing the insulation. A wood framed or equivalent baffle or retainer is required to be pro-</u>

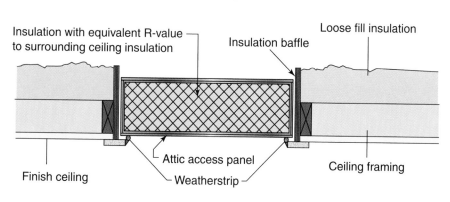

Attic access

vided when loose fill insulation is installed, the purpose of which is to prevent the loose fill insulation from spilling into the living space when the attic access is opened, and to provide a permanent means of maintaining the installed *R*-value of the loose fill insulation.

CHANGE SIGNIFICANCE: The new Section N1102.2.3 clarifies that doors or hatches providing access to unconditioned areas such as crawl spaces or attics are not exempt from the energy conservation requirements. Previously, the code did not specifically address these access openings, though the intent has always been that such openings meet the same requirements as other elements in the thermal envelope. These prescriptive provisions were proposed to encourage consistency in the application of the code. In addition to meeting the *R*-value of the surrounding ceiling, floor, or wall areas, the new section calls for weatherstripping of the hatch to reduce air leakage and infiltration. As is common in current construction practices, the code also now provides for a baffle or barrier to retain loose fill insulation at the access opening and to maintain the installed *R*-value of the insulation.

Section N1102.2.1 permits reduced ceiling *R*-values in certain Climate Zones when energy-type trusses are employed to maintain the required insulation thickness above exterior wall plates, providing a more energy-efficient assembly. Section N1102.2.2 recognizes the difficulty in obtaining the required *R*-values between rafters in limited areas of vaulted ceilings and permits a reduction in *R*-value for a ceiling area not greater than 500 square feet. The added text in these sections clarifies that such reductions apply only to the prescriptive values in Table N1102.1.

Both the *U*-factor alternative and the total UA alternative allow for the proper calculation of ceiling *U*-factors and permit trade-offs to allow for reduced ceiling insulation. The *U*-factor and UA approaches are based on the actual envelope component construction and do not utilize the adjustment factors needed for a practical *R*-value approach. The new language intends to reduce confusion and improve the consistency of application for these provisions.

Table N1102.2.5

Steel Framed Wall Insulation

CHANGE TYPE: Modification

CHANGE SUMMARY: Table N1102.2.10 now contains an option for a cold-formed steel framed wall without cavity insulation in Climate Zones 1 through 4 (except Marine). Continuous insulation rated not less than R-10 applied to a steel framed wall is considered equivalent to a wood framed wall with a cavity insulation *R*-value of R-13.

2009 CODE:
TABLE ~~N1102.2.4~~ <u>N1102.2.5</u> Steel Framed Ceiling, Wall, and Floor Insulation (*R*-Value)

Wood Frame *R*-value Requirement	Cold-Formed Steel Equivalent *R*-value[a]
Steel Framed Wall	
R-13	R-13+5 or R-15+4 or R-21+3 or <u>R-0+10</u>
R-19	R-13+9 or R-19+8 or R-25+7
R-21	R-13+10 or R-19+9 or R-25+8

(No change to remainder of table or footnotes.)

CHANGE SIGNIFICANCE: For all Climate Zones, Table N1102.2.5 previously required a combination of cavity insulation and continuous insulation for steel framed walls. The table indicates the cavity insulation *R*-value first followed by the continuous insulation *R*-value. Continuous insulation provides a thermal break at the studs and reduces air infiltration, resulting in a higher total wall *R*-value than an equivalent rating of cavity insulation. Recognizing the efficiency of continuous insulated sheathing and based on an application that already is used in the field, the code now offers an alternative method

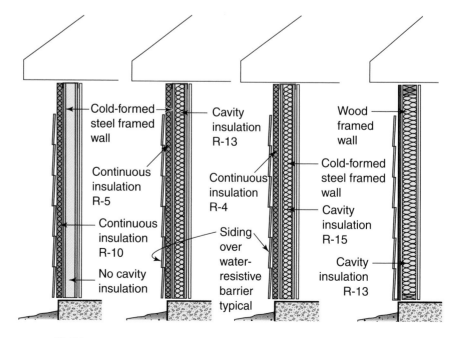

Wall insulation options in climate zones 1 through 4 (except marine)

employing only continuous sheathing for walls in Climate Zones 1 through 4 (except Marine). When applied to steel framed walls without cavity insulation, continuous insulated sheathing rated at R-10 is considered equivalent to wood framed walls with cavity insulation rated at R-13.

N1102.4.1

Sealing of the Building Thermal Envelope

CHANGE TYPE: Clarification

CHANGE SUMMARY: Attic access openings and rim joist junctions have been added to the list of specific locations requiring sealing to prevent air infiltration.

2009 CODE: N1102.4.1 Building Thermal Envelope. The building thermal envelope shall be durably sealed to limit infiltration. The sealing methods between dissimilar materials shall allow for differential expansion and contraction. The following shall be caulked, gasketed, weatherstripped or otherwise sealed with an air barrier material, suitable film or solid material.

1. All joints, seams and penetrations.
2. Site-built windows, doors, and skylights.
3. Openings between window and door assemblies and their respective jambs and framing.
4. Utility penetrations.
5. Dropped ceilings or chases adjacent to the thermal envelope.
6. Knee walls.
7. Walls and ceilings separating the garage from conditioned spaces.
8. Behind tubs and showers on exterior walls.
9. Common walls between dwelling units.
10. <u>Attic access openings.</u>
11. <u>Rim joist junction.</u>
~~10~~ <u>12.</u> Other sources of infiltration.

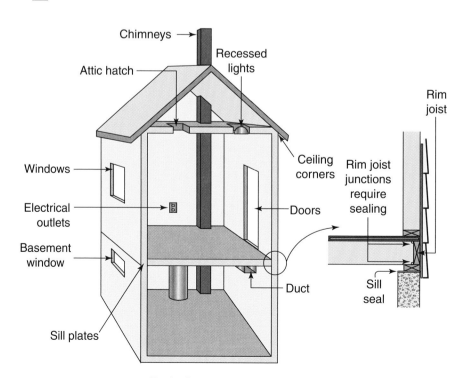

Typical sources of air leakage

CHANGE SIGNIFICANCE: Attic access openings and rim joist junctions with plates and other building materials have been identified as potential sources of significant air infiltration, thereby reducing the energy efficiency of the building thermal envelope. Though in principle these locations require sealing under the general provisions pointing to joints, seams, penetrations, and other sources of infiltration, specifically identifying access openings and rim joist junctions intends to provide clarification to both the contractor and inspector as to how to achieve code compliance. The expanded list promotes consistency in application of the thermal envelope sealing provisions.

N1102.4.5
Recessed Lighting

CHANGE TYPE: Modification

CHANGE SUMMARY: The IRC no longer recognizes multiple options for sealing recessed luminaires in the building thermal envelope. Recessed luminaires now must be insulation-contact (IC) rated and labeled as meeting the test criteria for air movement. The code also clarifies that these lighting fixtures must be sealed where they penetrate the finished wall or ceiling material.

2009 CODE: N1102.4.5 Recessed Lighting.   Recessed luminaires installed in the building thermal envelope shall be sealed to limit air leakage between conditioned and unconditioned spaces. <u>All recessed luminaires shall be IC rated and labeled as meeting ASTM E 283 when tested at 1.57 psi (75 Pa) pressure differential with no more than 2.0 cfm (0.944 L/s) of air movement from the conditioned space to the ceiling cavity. All recessed luminaires shall be sealed with a gasket or caulk between the housing and the interior wall or ceiling covering.</u> ~~By being:~~

> ~~1. IC-rated and labeled with enclosures that are sealed or gasketed to prevent air leakage to the ceiling cavity or unconditioned space; or~~
>
> ~~2. IC-rated and labeled as meeting ASTM E 283 when tested at 1.57 pounds per square foot (75 Pa) pressure differential with no more than 2.0 cubic feet per minute (0.944 L/s) of air movement from the conditioned space to the ceiling cavity; or~~
>
> ~~3. Located inside an airtight sealed box with clearances of at least 0.5 inch (13 mm) from combustible material and 3 inches (76 mm) from insulation.~~

CHANGE SIGNIFICANCE: If not properly sealed, recessed luminaires (lighting fixtures) are a significant source of air movement and heat loss through the building thermal envelope, thereby increasing energy consumption. Previous editions of the IRC permitted several

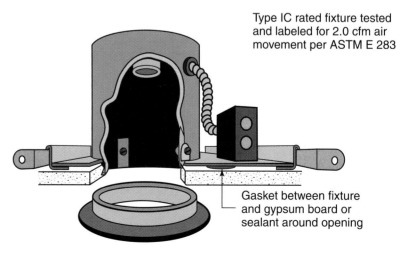

Type IC rated fixture tested and labeled for 2.0 cfm air movement per ASTM E 283

Gasket between fixture and gypsum board or sealant around opening

Recessed luminaire

options for sealing recessed luminaires installed in the building envelope. One option has been testing and labeling to meet air movement criteria. The 2009 IRC requires all recessed luminaires in the building envelope to be labeled as meeting the test criteria for resisting air movement through the fixture. Testing must conform to ASTM E283 *Standard Test Method for Determining Rate of Air Leakage Through Exterior Windows, Curtain Walls, and Doors Under Specified Pressure Differences Across the Specimen*. The code specifies the test criteria of 1.57 psi pressure differential with no more than 2.0 cfm of air movement from the conditioned space to the unconditioned space. The fixtures must also be insulation-contact (IC) rated, eliminating the need for an air space around the fixture housing for clearance to combustibles. The more stringent language aligns with the requirements of the International Energy Conservation Code (IECC). Tested and labeled fixtures have been available in the marketplace and have been required in some states for many years. Implementing a uniform standard for their installation in insulated assemblies will contribute to energy conservation efforts. These provisions are not intended to apply to recessed luminaires installed entirely within conditioned spaces of the building.

Inspections and building air leakage testing have determined that even when sealed luminaires are used, air leakage will occur if the fixture housing is not properly sealed to the wall or ceiling covering. For clarification, text has been added to emphasize the importance of installation practices that include sealing details. The code requires the installation of caulk or gasket material to fill the space between the fixture housing and the finish material.

N1103.7

Snow Melting System Controls

CHANGE TYPE: Addition

CHANGE SUMMARY: A new section in the IRC requires automatic shutoff controls on snow melt systems to conserve energy.

2009 CODE: <u>**N1103.7 Snow Melt System Controls.** Snow- and ice-melting systems, supplied through energy service to the building, shall include automatic controls capable of shutting off the system when the pavement temperature is above 50° F and no precipitation is falling and an automatic or manual control that will allow shutoff when the outdoor temperature is above 40° F.</u>

CHANGE SIGNIFICANCE: Particularly in areas of the country with high amounts of snowfall, such as mountainous regions of California, Colorado, Nevada, Utah, and Wyoming, snow melting systems are being installed with greater frequency on residential properties. Such systems employ approved hot water tubing or electric resistance wiring embedded in the concrete slab to clear snow and ice from driveways and sidewalks. Snow melting systems eliminate the need for snow removal by equipment or chemical means and provide greater safety for pedestrians and vehicles.

Previously, the energy provisions in the code pertained to only elements of the building. However, energy use for snow melting can be twice the energy use per square foot for a building. To further conserve energy, this new section in the code requires a snow detector that acti-

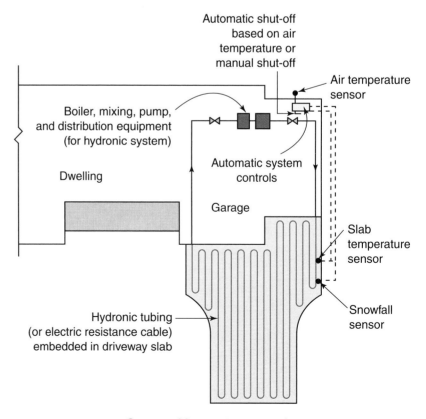

Snow melting system controls

vates the system from the idle mode to the snow melt mode when snow begins to fall and requires a slab temperature sensor that automatically turns the system off when the slab surface temperature is above 50 degrees Fahrenheit. In addition, the 2009 IRC requires either an automatic temperature control that shuts the system down when the outdoor temperature is above 40 degrees Fahrenheit or a manual control capable of shutting the system down. The new provisions are based on ANSI/ASHRAE/IESNA Standard 90.1 Section 6.4.3.8, Freeze Protection and Snow/Ice Melting Systems. This code change does not restrict the use or sizing of a snow melting system but does require controls so the system will operate more efficiently. The automatic controls conserve energy by activating the system only when necessary. Automatic start-up when light snow begins to fall allows adequate warm-up before a heavy snowfall. Manual systems typically use significantly more energy.

N1103.8
Swimming Pools

CHANGE TYPE: Addition

CHANGE SUMMARY: For energy conservation purposes, swimming pool heaters require automatic and manual on-off controls. Gas-fired pool heaters must be equipped with ignition means other than continuous pilot lights. The IRC now requires pool covers for all heated pools.

2009 CODE: **N1103.8 Pools.**   Pools shall be provided with energy-conserving measures in accordance with Sections N1103.8.1 through N1103.8.3.

N1103.8.1 Pool Heaters.   All pool heaters shall be equipped with a readily accessible on-off switch to allow shutting off the heater without adjusting the thermostat setting. Pool heaters fired by natural gas or LPG shall not have continuously burning pilot lights.

N1103.8.2 Time Switches.   Time switches that can automatically turn off and on heaters and pumps according to a preset schedule shall be installed on swimming pool heaters and pumps.

Exceptions:
1. Where public health standards require 24-hour pump operation
2. Where pumps are required to operate solar- and waste-heat-recovery pool heating systems

N1103.8.3 Pool Covers.   Heated pools shall be equipped with a vapor-retardant pool cover on or at the water surface. Pools heated to more than 90°F (32°C) shall have a pool cover with a minimum insulation value of R-12.

CHANGE SIGNIFICANCE: This change introduces new energy conservation requirements for swimming pools that are consistent with provisions in the IECC and ANSI/ASHRAE 90.1-2004, *Energy*

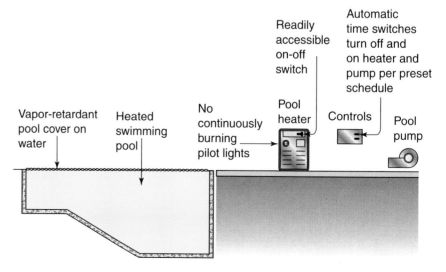

Energy conserving measures for heated pools

Standards for Buildings Except Low-Rise Residential Buildings. The first of the energy-saving measures requires a readily accessible on-off switch for pool heaters, allowing the user to manually reduce energy consumption. The code also no longer permits continuously operating pilot lights on gas-fired pool heaters. Automatic time switches are required on all pool heaters and pool filter pumps. The user will now have the ability to preset the timer to operate the pool filter pump for two to three hours per day rather than running the pump continuously, thereby realizing significant energy savings.

According to the U.S. Department of Energy (*Energy Efficiency and Renewable Energy: A Consumer's Guide to Energy Efficiency and Renewable Energy*), swimming pool covers reduce pool water heating energy requirements by 50 to 70 percent, reduce the make-up water needed by 30 to 50 percent, and reduce the pool's chemical consumption by 35 to 60 percent. Pool covers also reduce the ventilation and dehumidification energy required to control interior moisture loads and reduce the moisture loads on building assemblies that enclose pools. The IRC now requires swimming pool covers resting on the surface of the water for all heated swimming pools. When the temperature of the pool water exceeds 90 degrees Fahrenheit, the pool cover must have an insulation rating of not less than R-12. As in the case of environmental comfort heating and cooling, the pool water temperature will be under the control of the homeowner or occupant, who will bear the responsibility to provide the appropriate insulated cover when required by the code.

N1104

Lighting Systems

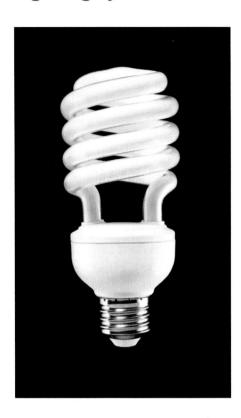

CHANGE TYPE: Addition

CHANGE SUMMARY: To conserve energy, the code now requires at least 50 percent of the lamps in permanently installed lighting fixtures to be compact fluorescent lamps or other high-efficacy lamps. A definition of *high-efficacy lamps* has been added to Section R202.

2009 CODE:
Section N1104
Lighting Systems
N1104.1 Lighting Equipment.   A minimum of fifty percent of the lamps in permanently installed lighting fixtures shall be high-efficacy lamps.

Section R202 Definitions
　　High-Efficacy Lamps.   Compact fluorescent lamps, T-8 or smaller diameter linear fluorescent lamps, or lamps with a minimum efficacy of:
　　1. 60 lumens per watt for lamps over 40 watts.
　　2. 50 lumens per watt for lamps over 15 watts to 40 watts.
　　3. 40 lumens per watt for lamps 15 watts or less.

CHANGE SIGNIFICANCE: With this new section in the 2009 IRC, the code initiates steps toward the goal of eliminating the use of incandescent lamps for permanent lighting fixtures as mandated by the energy bill by the year 2012. The new provisions require at least 50 percent of permanent lighting fixtures (*luminaires* in the IRC electrical provisions) to be equipped with high-efficacy lamps. One option for satisfying the definition of high-efficacy lamps and for conserving energy is the *compact fluorescent lamp* (CFL). Lighting, primarily by incandescent lamps, currently accounts for approximately 12 percent of primary residential energy consumption. CFLs use about 80 percent less energy and last 6 to 10 times longer than standard incandescent lighting. Assuming a cost of $1.50 per bulb, electricity at 9 cents per kwh, and an average one half hour per day of use for each light, the payback time is less than two years. Many lights are used for more than an hour per day, yielding paybacks of less than a year. Limiting this requirement to 50 percent of the permanent light fixtures in a residence ensures that there will be plenty of exceptions for situations where a CFL might not work as well, such as dimmable fixtures.

PART 5

Mechanical

Chapters 12 through 23

The IRC, as a code specific to the entire construction of residential buildings that fall under its scope, contains provisions for mechanical, fuel gas, plumbing, and electrical systems of the building. These systems are covered in various parts of the IRC, beginning with Part 5.

This part contains administrative provisions unique to the application and enforcement of mechanical systems, as well as the technical provisions related to the design and installation of mechanical systems. Chapter 13 provides the general requirements for all mechanical systems and addresses the labeling of appliances, types of fuel used, access to appliances for repair and maintenance, and other issues such as clearances from combustibles. The remainder of Part 5 deals with issues of specific mechanical systems such as heating and cooling systems, exhaust systems, ducts, boilers, and hydronic piping. Part 5 also contains two chapters specific to oil tanks, oil pumps, and solar systems. ■

M1305.1.4.1 AND M1308.3

Ground Clearance for Appliances

M1307.3.1

Protection from Impact

M1502

Clothes Dryer Exhaust

M1503.4

Makeup Air for Kitchen Exhaust Hoods

M1601.3 AND M1601.4

Duct Insulation Materials and Duct Installation

M1601.6

Independent Garage HVAC Systems

M1602.2

Prohibited Sources of Outdoor and Return Air

M1701

Combustion Air

M2103.2

Hydronic Floor Heating Systems

M2104 AND TABLE M2101.1

Hydronic Piping Materials

M1305.1.4.1 and M1308.3

Ground Clearance for Appliances

CHANGE TYPE: Modification

CHANGE SUMMARY: The ground clearance requirements for mechanical equipment and appliances have been consolidated into one section.

2009 CODE: M1305.1.4.1 Ground Clearance. <u>Equipment and appliances supported from the ground shall be level and firmly supported on a concrete slab or other approved material extending not less than 3 inches (76 mm) above the adjoining ground. Such support shall be in accordance with the manufacturer's installation instructions.</u> Appliances suspended from the floor shall have a clearance of not less than 6 inches (152 mm) from the ground.

~~M1308.3 Foundations and Supports.~~ ~~Foundations and supports for outdoor mechanical systems shall be raised at least 3 inches (76 mm) above the finished grade, and shall also conform to the manufacturer's installation instructions.~~

CHANGE SIGNIFICANCE: Section M1305.1.4 sets requirements for the access to mechanical appliances, other than gas-fired equipment, installed in underfloor spaces, and Section M1305.1.4.1 prescribes the minimum clearance above the ground for these appliances. When supported on a slab or approved pad, the support must extend above ground not less than 3 inches. This distance is increased to 6 inches for appliances suspended from the floor above the crawl space. In the 2006 IRC, the 3-inch clearance above finished grade for outdoor mechanical equipment was located in Section M1308.3, which further specified that the installation conform to the manufacturer's instructions. The 2009 IRC deletes Section M1308.3 and consolidates

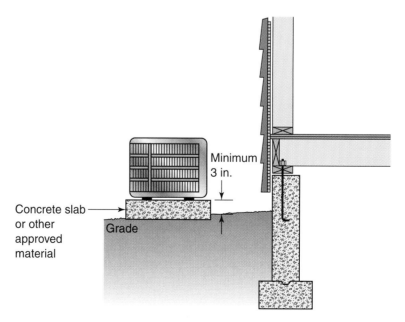

Mechanical appliance installed at least 3 in. above adjoining ground

all ground clearance requirements in Section M1305.1.4.1. The code also adds the stipulation to this section that support must satisfy the manufacturer's installation requirements. The intent of this change is that Section M1305.1.4.1 now applies to mechanical equipment and appliances in both crawl space and outdoor locations to protect the units from moisture and corrosion.

The IRC fuel gas provisions regulate the installation of gas-fired appliances, including support and ground clearance requirements. For appliances installed outdoors or in underfloor spaces, Section G2408.4 requires only that concrete pads or other approved material extend above the ground and be level. For suspended equipment and appliances, this section requires a 6-inch ground clearance and matches the provisions in Section M1305.1.4.1.

M1307.3.1

Protection from Impact

CHANGE TYPE: Modification

CHANGE SUMMARY: The 2009 IRC expands the requirements for protecting appliances from vehicle impact damage to include locations other than garages and carports.

2009 CODE: M1307.3.1 Protection from Impact. Appliances ~~located in a garage or carport shall be protected from impact by automobiles~~ <u>shall not be installed in a location subject to vehicle damage except where protected by approved barriers.</u>

CHANGE SIGNIFICANCE: Previously, the impact protection requirements applied to only mechanical appliances installed in garages or carports. While such installations are common and are most likely to be affected by close proximity to cars and trucks, outdoor locations, particularly those adjacent to driveways, also may be vulnerable to vehicle impact. The 2009 IRC is more inclusive in recognizing that mechanical appliances installed in any location that is subject to damage from vehicles need to be protected with an approved barrier. Barriers may be steel pipe or concrete bollards, curbs or any structure that affords protection to the appliance.

Note that suspended appliances with sufficient clearance above the floor and appliances installed in an alcove out of the path of vehicle travel are not subject to impact and do not require additional barriers. The new language in Section M1307.3.1 is similar to that of Section 303.4 of the International Mechanical Code (IMC).

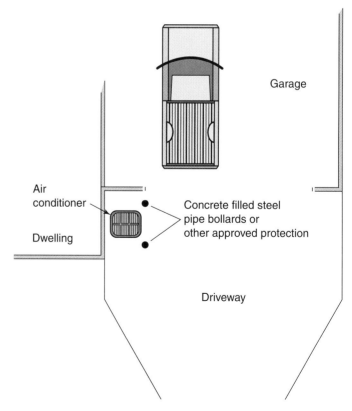

Protection from vehicle impact damage

M1502
Clothes Dryer Exhaust

CHANGE TYPE: Modification

CHANGE SUMMARY: Dryer exhaust duct installation under the 2009 IRC focuses primarily on the dryer manufacturer's installation instructions. The code clarifies the provisions for duct materials and installation to reflect current industry practices. Equivalent lengths for fittings appear in a new table and are based on the radius and type of fitting. When a concealed exhaust system with a length greater than 25 feet is installed in accordance with the dryer manufacturer's installation instructions, the developed length must be identified with a permanent marker. New provisions require protection of the dryer duct against penetration by drywall fasteners.

2009 CODE: M1502.1 General. <u>Clothes dryers shall be exhausted in accordance with the manufacturer's instructions.</u>

M1502.2 Independent Exhaust Systems. Dryer exhaust systems shall be independent of all other systems and shall convey the moisture to the outdoors.

> **Exception:** This section shall not apply to listed and labeled condensing (ductless) clothes dryers.

~~**M1502.2**~~ **M1502.3 Duct Termination.** Exhaust ducts shall terminate on the outside of the building. Exhaust duct terminations shall be in accordance with the dryer manufacturer's installation instructions. <u>If the manufacturer's instructions do not specify a termination location, the</u> exhaust ducts shall terminate not less than 3 feet (914 mm) in any direction from openings into buildings. Exhaust duct terminations shall be equipped with a backdraft damper. Screens shall not be installed at the duct termination.

M1502.4 Dryer Exhaust Ducts. <u>Dryer exhaust ducts shall conform to the requirements of Sections M1502.4.1 through M1502.4.6.</u>

M1502 continues

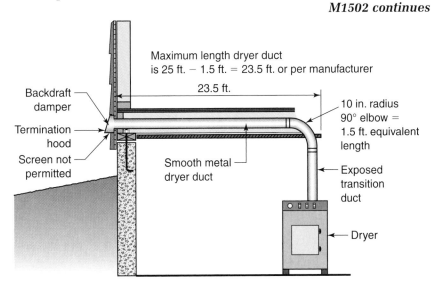

Maximum length dryer duct
is 25 ft. − 1.5 ft. = 23.5 ft. or per manufacturer

23.5 ft.

Backdraft damper

Termination hood

Screen not permitted

Smooth metal dryer duct

10 in. radius 90° elbow = 1.5 ft. equivalent length

Exposed transition duct

Dryer

Clothes dryer exhaust

M1502 continued

M1502.3 Duct Size. ~~The diameter of the exhaust duct shall be as required by the clothes dryer's listing and the manufacturer's installation instructions.~~

M1502.4.1 Material and Size. Exhaust ducts shall have a smooth interior finish and shall be constructed of metal a minimum 0.016-inch (0.4 mm) thick. The exhaust duct size shall be 4 inches (102 mm) nominal in diameter.

M1502.4.2 Duct Installation. Exhaust ducts shall be supported at 4 foot (1219 mm) intervals and secured in place. The insert end of the duct shall extend into the adjoining duct or fitting in the direction of airflow. Ducts shall not be joined with screws or similar fasteners that protrude into the inside of the duct.

~~M1502.4~~ M1502.4.3 Transition Ducts. Transition ducts used to connect the dryer to the exhaust duct system shall be a single length that is listed and labeled in accordance with UL 2158A. Transition ducts shall be a maximum of 8 feet (2438 mm) in length. Transition ducts shall not be concealed within construction. ~~Flexible transition ducts used to connect the dryer to the exhaust duct system shall be limited to single lengths, not to exceed 8 feet (2438 mm) and shall be listed and labeled in accordance with UL 2158A.~~

M1502.4.4 Duct Length. The maximum allowable exhaust duct length shall be determined by one of the methods specified in Section M1502.4.4.1 or M1502.4.4.2.

M1502.4.4.1 Specified Length. The maximum length of the exhaust duct shall be 25 feet (7620 mm) from the connection to the transition duct from the dryer to the outlet terminal. Where fittings are used, the maximum length of the exhaust duct shall be reduced in accordance with Table M1502.4.4.1.

M1502.4.4.2 Manufacturer's Instructions. The size and maximum length of the exhaust duct shall be determined by the dryer manufacturer's installation instructions. The code official shall be provided with a copy of the installation instructions for the make and model of the dryer at the concealment inspection. In the absence of fitting equivalent length calculations from the clothes dryer manufacturer, Table M1502.4.4.1 shall be used.

M1502.4.5 Length Identification. Where the exhaust duct is concealed within the building construction, the equivalent length of the exhaust duct shall be identified on a permanent label or tag. The label or tag shall be located within 6 feet (1829 mm) of the exhaust duct connection.

M1502.4.6 Exhaust Duct Required. Where space for a clothes dryer is provided, an exhaust duct system shall be installed. Where the clothes dryer is not installed at the time of occupancy the exhaust duct shall be capped or plugged in the space in which it originates and identified and marked "future use."

Exception: Where a listed condensing clothes dryer is installed prior to occupancy of the structure.

~~**M1502.5 Duct Construction.** Exhaust ducts shall be constructed of minimum 0.016-inch-thick (0.4 mm) rigid metal ducts, having smooth interior surfaces with joints running in the direction of air flow. Exhaust ducts shall not be connected with sheet-metal screws or fastening means which extend into the duct.~~

~~**M1502.6 Duct Length.** The maximum length of a clothes dryer exhaust duct shall not exceed 25 feet (7620 mm) from the dryer location to the wall or roof termination. The maximum length of the duct shall be reduced 2.5 feet (762 mm) for each 45-degree (0.8 rad) bend and 5 feet (1524 mm) for each 90-degree (1.6 rad) bend. The maximum length of the exhaust duct does not include the transition duct.~~

Exceptions:
~~1. Where the make and model of the clothes dryer to be installed is known and the manufacturer's installation instructions for the dryer are provided to the building official, the maximum length of the exhaust duct, including any transition duct, shall be permitted to be in accordance with the dryer manufacturer's installation instructions.~~
~~2. Where large-radius 45-degree (0.8 rad) and 90-degree (1.6 rad) bends are installed, determination of the equivalent length of clothes dryer exhaust duct for each bend by engineering calculation in accordance with the ASHRAE Fundamentals Handbook shall be permitted.~~

M1502.5 Protection Required. Protective shield plates shall be placed where nails or screws from finish or other work are likely to penetrate the clothes dryer exhaust duct. Shield plates shall be placed on the finished face of all framing members where there is less than 1¼ inches (32 mm) between the duct and the finished face of the framing member. Protective shield plates shall be constructed of steel, shall have a thickness of 0.062 inch (1.6 mm), and shall extend a minimum of 2 inches above sole plates and below top plates.

CHANGE SIGNIFICANCE: The modification to Section M1502.1 and M1502.3 emphasizes that the manufacturer's installation instructions

M1502 continues

TABLE M1502.4.4.1 **Dryer Exhaust Duct Fitting Equivalent Length**

Dryer Exhaust Duct Fitting Type	Equivalent Length (feet)
4 inches radius mitered 45 degree elbow	2 feet 6 inches
4 inches radius mitered 90 degree elbow	5 feet
6 inches radius smooth 45 degree elbow	1 foot
6 inches radius smooth 90 degree elbow	1 foot 9 inches
8 inches radius smooth 45 degree elbow	1 foot
8 inches radius smooth 90 degree elbow	1 foot 7 inches
10 inches radius smooth 45 degree elbow	9 inches
10 inches radius smooth 90 degree elbow	1 foot 6 inches

M1502 continued

are the first source for dryer exhaust installation and termination requirements, which are related to the design and testing of the specific model of dryer. For example, the manufacturer may permit a termination location with clearance to openings less than 3 feet or may require a termination point greater than 3 feet from openings. The prescriptive requirement for the minimum 3-foot distance between dryer exhaust terminations and openings into buildings applies only when the clearance is not specified by the manufacturer or is not known.

The dryer manufacturer's installation instructions also govern the maximum developed length of the exhaust duct, including provisions for fittings, but only if the model of dryer is known and the installation instructions are submitted to the building official. Modern dryers are increasingly efficient and are generally designed to exhaust greater distances than otherwise allowed by the code, with some models permitting duct lengths up to 90 feet and many models reaching 60 feet. When a dryer exhaust duct is concealed, the code now requires a permanent sign, label, or tag identifying the developed length of the exhaust duct. This new requirement intends to alert homeowners installing replacement dryers to match the specifications for the make and model to the existing exhaust duct installation.

Elbow fittings increase the resistance to air flow and reduce the allowable length of exhaust duct. Previously, the code required a reduction of 2 feet 6 inches for 45-degree elbows and 5 feet for 90-degree elbows. These deductions were based on 4-inch radius fittings. An exception referenced the *ASHRAE Fundamentals Handbook* for large-radius fittings, but this required calculations based on air friction resistance. To consolidate the information in the code for ease of use, the 2009 IRC places the reductions for fittings in a new table that includes both 4-inch radius fittings and greater radius long-sweep fittings. The equivalent lengths of elbow fittings are based on values published by the American Society of Heating, Refrigerating and Air-Conditioning Engineers (ASHRAE) and Sheet Metal and Air Conditioning Contractors' National Association (SMACNA). Testing at Underwriters Laboratories (UL) verified that the 10-inch-radius elbows perform significantly better than 4-inch radius elbows.

The new Section M1502.4.6 requires installation of a dryer exhaust duct system at the time of rough-in and is similar to language in Section G2439.5.2 of the 2006 IRC for gas dryers. Most dwellings have a space for a dryer installation, and the intent of this language is to require a clothes dryer exhaust duct system to be installed at the time of construction so it may be inspected for compliance with the code. With the addition of Section M15024.6, the requirement applies whether the space is set up for an electric dryer or a gas-fired dryer.

The 2009 IRC also adds provisions for protecting dryer ducts from penetration by fasteners. The new requirements are similar to protection requirements for piping and gas vents. In the case of dryer ducts, the concern of fastener penetration is related to the buildup of lint catching on the penetrating fasteners over time and increasing the fire hazard.

CHANGE TYPE: Addition

CHANGE SUMMARY: A new Section M1503.4 requires makeup air for kitchen exhaust hoods with a rating greater than 400 cfm. The makeup air system must be synchronized with the operation of the exhaust hood.

2009 CODE: **M1503.4 Makeup Air Required.** Exhaust hood systems capable of exhausting in excess of 400 cubic feet per minute (0.19 m³/s) shall be provided with makeup air at a rate approximately equal to the exhaust air rate. Such makeup air systems shall be equipped with a means of closure and shall be automatically controlled to start and operate simultaneously with the exhaust system.

CHANGE SIGNIFICANCE: This change is a result of concerns that residential kitchens are becoming larger, and installation of kitchen exhaust hoods with high-velocity fans are increasing. With tighter building thermal envelopes, the high rate of exhaust requires outside makeup air to prevent negative pressure and the accompanying adverse effects on other mechanical appliances and systems. Residential kitchens have traditionally been equipped with 400 cfm or less exhaust systems that have performed satisfactorily, and such systems are exempt from the new makeup air provisions. However, kitchen hoods exceeding 400 cfm are common in the marketplace, with some models reaching a capacity of 1300 cfm or greater. The new provisions specifically address the need for makeup air for high-velocity systems and prescribe an amount approximately equal to the rate of exhaust. The makeup air system requires a motorized damper that is interlocked with the kitchen range hood exhaust system to prevent air from entering the building when the hood is not in operation.

M1503.4
Makeup Air for Kitchen Exhaust Hoods

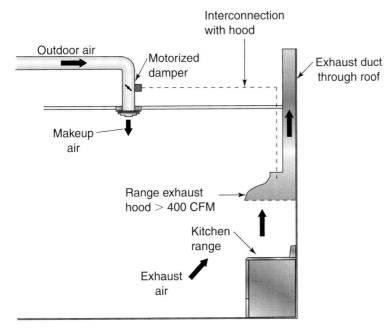

Required makeup air for kitchen exhaust hoods exceeding 400 CFM

M1601.3 and M1601.4

Duct Insulation Materials and Duct Installation

CHANGE TYPE: Modification

CHANGE SUMMARY: The code adds an alternative testing method for determining flame spread and smoke-developed indices of duct insulation materials. Spray-applied polyurethane foam insulation is now recognized for the insulation and sealing of ducts in specific attic and crawl space applications. Changes to the provisions for connecting and sealing ducts clarify their application.

2009 CODE: ~~**M1601.2.1**~~ **M1601.3 Duct Insulation Materials.** Duct insulation materials shall conform to the following requirements:

1. Duct coverings and linings, including adhesives where used, shall have a flame spread index not higher than 25 and a smoke-developed index not over 50 when tested in accordance with ASTM E 84 <u>or UL 723</u>, using the specimen preparation and mounting procedures of ASTM E 2231.

   <u>**Exception:** Spray application of polyurethane foam to the exterior of ducts in attics and crawl spaces shall be permitted subject to all of the following:</u>

   <u>1.</u> <u>The flame spread index is not greater than 25 and the smoke-developed index is not greater than 450 at the specified installed thickness.</u>

   <u>2.</u> <u>The foam plastic is protected in accordance with the ignition barrier requirements of Sections R316.5.3 and R316.5.4.</u>

   <u>3.</u> <u>The foam plastic complies with the requirements of Section R316.</u>

2. (No change to text.)

3. External duct insulation and factory-insulated flexible ducts shall be legibly printed or identified at intervals not longer than 36 inches (914 mm) with the name of the manufacturer;

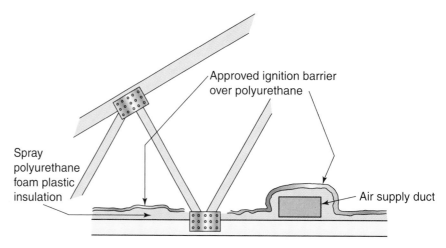

Spray polyurethane attic insulation approved for sealing and insulating ductwork

the thermal resistance R-value at the specified installed thickness; and the flame spread and smoke-developed indices of the composite materials. <u>Spray polyurethane foam manufacturers shall provide the same product information and properties, at the nominal installed thickness, to the customer in writing, at the time of foam application.</u> All duct insulation product R-values shall be based on insulation only, excluding air films, vapor retarders, or other duct components, and shall be based on tested C-values at 75°F (24°C) mean temperature at the installed thickness, in accordance with recognized industry procedures. The installed thickness of duct insulation used to determine its R-value shall be determined as follows:

3.1 (through) 3.3. (No change to text.)

3.4. <u>For spray polyurethane foam, the aged R-value per inch measured in accordance with recognized industry standards shall be provided to the customer in writing at the time of foam application. In addition, the total R-value for the nominal application thickness shall be provided.</u>

~~M1601.3~~ <u>M1601.4</u> Installation. Duct installation shall comply with Sections M1601.4.1 through M1601.4.6.

~~M1601.3.1~~ <u>M1601.4.1</u> Joints and Seams. Joints of duct systems shall be made substantially airtight by means of tapes, mastics, <u>liquid sealants,</u> gasketing or other approved closure systems. Closure systems used with rigid fibrous glass ducts shall comply with UL 181A and shall be marked "181A-P" for pressure-sensitive tape, "181A-M"for mastic, or "181 A-H" for heat-sensitive tape. Closure systems used with flexible air ducts and flexible air connectors shall comply with UL 181B and shall be marked "181B-FX" for pressure-sensitive tape or "181B-M" for mastic. Duct connections to flanges of air distribution system equipment or sheet metal fittings shall be mechanically fastened. Mechanical fasteners for use with flexible nonmetallic air ducts shall comply with UL 181B and shall be marked "181B-C." Crimp joints for round metal ducts shall have a contact lap of at least 1½ inches (38 mm) and shall be mechanically fastened by means of at least three sheet-metal screws or rivets equally spaced around the joint. <u>Closure systems used to seal metal ductwork shall be installed in accordance with the manufacturer's installation instructions.</u>

Exceptions:
1. <u>Application of spray polyurethane foam shall be permitted without additional joint seals.</u>
2. <u>Where a duct connection is made that is partially inaccessible, three screws or rivets shall be equally spaced on the exposed portion of the joint so as to prevent a hinge effect.</u>
3. <u>Continuously welded and locking type longitudinal joints and seams in ducts operating at static pressures less than 2 inches of water column (500 Pa) pressure classification shall not require additional closure systems.</u>

M1601.3 and M1601.4 continues

M1601.3 and M1601.4 continued

M1601.3.4 M1601.4.5 Duct Insulation. Duct insulation shall be installed in accordance with the following requirements:

1. A vapor retarder having a maximum permeance of 0.05 perm [2.87 ng/(s · m^2 · Pa)] in accordance with ASTM E 96, or aluminum foil with a minimum thickness of 2 mils (0.05 mm), shall be installed on the exterior of insulation on cooling supply ducts that pass through nonconditioned spaces conducive to condensation <u>except where the insulation is spray polyurethane foam with a maximum water vapor permeance of 3 perm per inch [1722 ng/(s · m^2 · Pa)] at the installed thickness.</u>

2. **through 3.** [No change to text]

CHANGE SIGNIFICANCE: This change adds a direct reference to UL 723 *Standard for Test for Surface Burning Characteristics of Building Materials* as an alternative to ASTM E 84 *Standard Test Method for Surface Burning Characteristics of Building Materials* for testing of duct coverings and linings. These two standards describe the same test method and specify identical test apparatus and test procedures. The purpose of the tests is to determine the comparative burning characteristics of the material under test by evaluating the spread of flame over its surface and the density of the smoke developed when exposed to a test fire, and thus to establish a basis on which surface burning characteristics of different materials are compared. The inclusion of this alternate test method provides the building official with the flexibility to accept listed and labeled products evaluated in accordance with ASTM E 84 or UL 723.

The IRC recognizes spray polyurethane foam as an approved insulation in attics and crawl spaces when protected by an ignition barrier. The added language in the 2009 edition clarifies that spray application of the polyurethane product is permitted to cover ducts installed in attics and crawl spaces, provided an ignition barrier is installed to comply with Section R316. In addition to providing continuous insulation and satisfying the energy-efficiency requirements for the thermal envelope, spray polyurethane provides air leakage control to the duct system from the duct exterior. The prescribed flame spread and smoke-developed indices are generally consistent with provisions in Chapter 3 of the IRC and Section 719.7 of the 2006 IBC, the latter permitting the use of exposed insulation and covering on pipe and tubing when the flame spread index is not more than 25 and the smoke-developed index is not more than 450. The requirement to provide product information and the *R*-value for the installed thickness of polyurethane insulation is similar to the certification requirements for thermal envelope insulation in Section N1101.4.

The 2009 IRC clarifies that spray polyurethane foam, when used as continuous insulation covering ducts in attics and crawl spaces, is satisfactory for the sealing of joints and seams in ductwork, and additional sealants are not required. Through field experience, the vapor permeability of spray polyurethane foam has been considered satisfactory in some applications without vapor retarders. The code now recognizes spray polyurethane foam with a maximum water vapor permeance of

3 perms per inch as an approved alternative vapor retarder for duct insulation in the above-mentioned attic and crawl space installations.

The referenced standards for joints and seams of ducts apply to factory-made rigid fibrous ducts and flexible air ducts and connectors. The tapes and mastics meeting the listings of these standards are not tested on metal ducts. With the addition of liquid sealants, the code clarifies the approved methods for sealing metal ducts and permits liquid sealants that are listed and labeled for the intended application. This change relies on terminology from the *HVAC Duct Construction Standards—Metal and Flexible* published by the Sheet Metal and Air Conditioning Contractors' National Association (SMACNA). The SMACNA standard specifically distinguishes between a liquid sealant and mastic, and recognizes the use of a liquid sealant as an adequate product for sealing ducts. Liquid sealants are widely available from several different manufacturers. In addition, the code clarifies that all closure systems must be in accordance with the manufacturer's installation instructions. The change intends to promote uniform application of the code provisions.

The duct connection provisions also have been expanded to recognize alternate fastener arrangement, welding- or locking-type closures, and joining methods for plastic ducts in accordance with the manufacturer's installation instructions. When placing round ductwork in the joist space or against framing members, there is often insufficient access to install three screws or rivets in an equally spaced manner around the joint. The new exception permits spacing the fasteners on the exposed portion of the joint but requires the installer to secure the joint in a manner that would prevent a hinging effect.

M1601.6

Independent Garage HVAC Systems

CHANGE TYPE: Addition

CHANGE SUMMARY: New language clarifies that an HVAC system is not permitted to serve both the dwelling unit and the garage.

2009 CODE: **M1601.6 Independent Garage HVAC Systems.** <u>Furnaces and air-handling systems that supply air to living spaces shall not supply air to or return air from a garage.</u>

CHANGE SIGNIFICANCE: Section R302.5.2 (R309.1.1 in the 2006 IRC) regulates ducts located in a garage and penetrating the dwelling/garage separation. The IRC does not permit such ducts to have openings into the garage. The intent is to reduce the likelihood of contaminants such as carbon monoxide, flammable vapors, or noxious odors migrating from the garage to the dwelling unit through ducts. The application of this requirement was not always clear. For example, since this section was addressing a penetration, it was possible to infer that opening protection such as a fire damper was permitted for a penetrating duct serving both the dwelling and the garage. The new section in the duct systems chapter intends to leave no doubt and specifically prohibits a furnace or air-handling unit duct system serving the living space to be used for heating or cooling a garage. The garage is also a prohibited location for drawing return air to the dwelling HVAC system. A completely independent system, such as a unit space heater, is required for conditioning of the garage space.

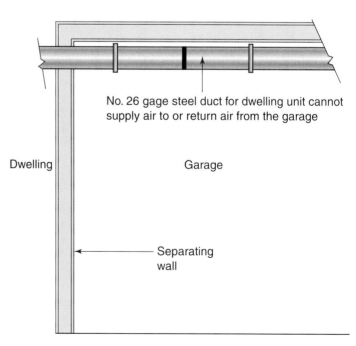

No. 26 gage steel duct for dwelling unit cannot supply air to or return air from the garage

Dwelling

Garage

Separating wall

Separate HVAC systems for dwellings and garages

M1602.2

Prohibited Sources of Outdoor and Return Air

CHANGE TYPE: Modification

CHANGE SUMMARY: Unconditioned attics and crawl spaces are now specifically prohibited as sources of outdoor or return air for HVAC systems.

2009 CODE: M1602.2 Prohibited Sources.   Outdoor and return air for a forced-air heating or cooling system shall not be taken from the following locations:

1. (through) **3.** (No change to text.)
4. A closet, bathroom, toilet room, kitchen, garage, mechanical room, <u>boiler room,</u> furnace room, <u>unconditioned attic</u> or other dwelling unit.
5. (No change to text.)
6. <u>An unconditioned crawl space by means of direct connection to the return side of a forced air system. Transfer openings in the crawl space enclosure shall not be prohibited.</u>

CHANGE SIGNIFICANCE: This change adds unconditioned attics and crawl spaces to the list of prohibited sources for obtaining outdoor or combustion air for forced-air heating and cooling systems. The new text intends to prevent mold and odors from being introduced into the conditioned space. Since conditioned spaces are less likely to promote mold and bacteria growth, taking return air from conditioned attics and crawl spaces is still acceptable. For gas-fired appliances, the provisions in Section G2407.6 for drawing combustion air from ventilated attics and crawl spaces are still in place and are not affected by the prohibition for outdoor and return air as specified in Section M1602.2.

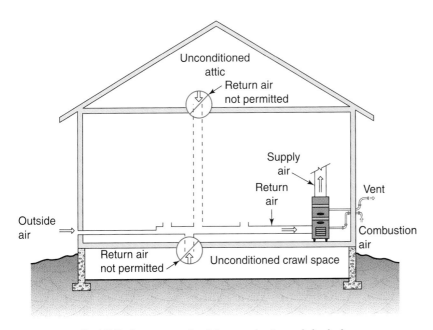

Prohibited sources of outdoor and return air include unconditioned attics and crawl spaces

M1701

Combustion Air

CHANGE TYPE: Modification

CHANGE SUMMARY: Chapter 17, *Combustion Air*, Sections M1701.1.1 through M1703.5, Figures M1702.2 through M1703.3, and three associated definitions of Section R202 have been deleted without substitution. The remaining Section M1701.1 references the combustion air requirements of the manufacturer for solid-fuel-burning appliances and of NFPA 31 for oil-fired appliances. Prescriptive methods for providing combustion air to gas-fired appliances are found in IRC Chapter 24.

2009 CODE: **M1701.1 ~~Air Supply~~ Scope.** ~~Liquid- and solid-fuel-burning appliances shall be provided with a supply of air for fuel combustion, draft hood dilution and ventilation of the space in which the appliance is installed, in accordance with Section M1702 or Section M1703.~~ Solid-fuel-burning appliances shall be provided with combus-

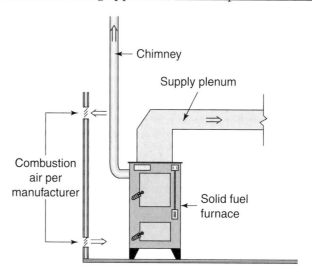

Combustion air for solid fuel appliances

tion air, in accordance with the appliance manufacturer's installation instructions. Oil-fired appliances shall be provided with combustion air in accordance with NFPA 31. The methods of providing combustion air in this chapter do not apply to fireplaces, fireplace stoves, and direct-vent appliances. The requirements for combustion and dilution air for gas-fired appliances shall be in accordance with Chapter 24.

M1701.2 Combustion Air Opening Location. In areas prone to flooding as established in Table R301.2(1), *combustion* openings shall be located at or above the elevation required in Section R322.2.1 or R322.3.2.

M1701.1.1 Buildings of Unusually Tight Construction.
M1701.2 Exhaust and Ventilation System.
M1701.3 Volume Dampers Prohibited.
M1701.4 Prohibited Sources.
M1701.5 Opening Area.
M1701.6 Opening Location.
Section M1702 All Air From Inside the Building
Section M1703 All Air from Outdoors
Section R202 Definitions
Confined Spaces. Unconfined Space.
Unusually Tight Construction.
Because this code change deleted substantial portions of Chapter 17, the entire code change text is too extensive to be included here. Refer to Code Change M108-06/07 Part II in the *2009 IRC Code Changes Resource Collection* for the complete text and history of the code change.

CHANGE SIGNIFICANCE: In the 2006 IRC, the combustion air provisions for solid and liquid fuel appliances were based on and similar to the corresponding requirements for gas-fired appliances in Chapter 24. This change recognizes that there are some inherent differences in the operation characteristics of the respective appliances and that combustion air provisions should be specific to the fuel type. As a result, the 2009 IRC references the manufacturer's installation instructions for solid-fuel-burning appliances in providing the appropriate quantities of combustion air. Similarly, the provisions of NFPA 31 *Standard for the Installation of Oil-Burning Equipment* now govern the combustion air requirements for residential oil-fired appliances. All other combustion air provisions of Chapter 17 have been deleted without substitution.

Previous to the 2006 IRC, *confined spaces, unconfined space,* and *unusually tight construction* described criteria for determining outdoor combustion air requirements based on available air volume and air infiltration. Subsequent testing by the fuel gas industry determined that confined or unconfined space and the method of construction were not significant factors for determining the need to obtain outdoor combustion air. Consequently, these terms were deleted from the fuel-gas provisions of Chapter 24 but remained in the general definitions Section R202 and still applied to the Chapter 17 provisions for solid

M1701 continues

M1701 continued   and liquid fuel appliances. With this change to Chapter 17, the terms *confined spaces, unconfined space,* and *unusually tight construction* no longer have an application in the 2009 IRC and have been deleted.

Section M1701.1 retains the text related to fireplaces, fireplace stoves, and direct-vent appliances, whose installation and combustion air requirements are governed by the manufacturer's installation instructions and the relevant provisions in IRC Chapter 10, *Chimneys and Fireplaces,* and Chapter 24, *Fuel Gas.*

M2103.2
Hydronic Floor Heating Systems

CHANGE TYPE: Addition

CHANGE SUMMARY: Hydronic radiant floor heating systems now require thermal insulation installed below the piping or tubing. The *R*-value marking on the insulation must be visible before finish materials are installed. A thermal break is required between a heated slab and the foundation.

2009 CODE: **M2103.2 Thermal Barrier Required.** Radiant floor heating systems shall be provided with a thermal barrier in accordance with Sections M2103.2.1 through M2103.2.4.

M2103.2.1 Slab on Grade Installation. Radiant piping used in slab-on-grade applications shall have insulating materials having a minimum *R*-value of 5 installed beneath the piping.

M2103.2.2 Suspended Floor Installation. In suspended floor applications, insulation shall be installed in the joist bay cavity serving the heating space above and shall consist of materials having a minimum *R*-value of 11.

M2103.2 continues

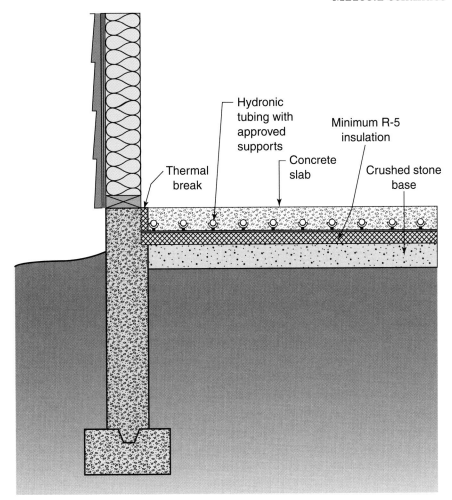

Hydronic floor heating system

M2103.2 continued

M2103.2.3 Thermal Break Required. A thermal break consisting of asphalt expansion joint materials or similar insulating materials shall be provided at a point where a heated slab meets a foundation wall or other conductive slab.

M2103.2.4 Thermal Barrier Material Marking. Insulating materials utilized in thermal barriers shall be installed such that the manufacturer's *R*-value mark is readily observable upon inspection.

Exception: Insulation shall not be required in engineered systems where it can be demonstrated that the insulation will decrease the efficiency or have a negative effect on the installation.

CHANGE SIGNIFICANCE: The 2009 IRC contains thermal insulation requirements for hydronic radiant floor heating systems to improve the effectiveness and efficiency of the system and to conserve energy. Efficient operation of a radiant system depends on thermal insulation installed below the hydronic tubing to direct the heat to the conditioned space above. In the case of a slab on grade application without insulation, the ground requires a substantial charging of energy before hitting a point of equilibrium where the majority of heat rises as intended. This flow of thermal energy to warm the underlying ground affects the overall energy consumption and can negatively affect the performance of the system. In much the same way, the lack of insulation below the hydronic tubing installed in or on the floor/ceiling assembly results in an unbalanced system directing disproportionate amounts of heat to the space below and inadequate energy to the space above as intended. The boiler or water heater manufacturer's installation instructions do not typically address installation specifications for the distribution system or insulation requirements, and this change intends to fill a gap in the hydronic piping provisions and supplement the energy-efficiency provisions of Chapter 11. The minimum *R*-values of R-5 for slab on grade insulation and R-11 for the floor/ceiling assembly insulation are consistent with provisions in IRC Section N1102. For the inspector to verify insulation requirements, the manufacturer's identification of the *R*-value must be visible before finish materials are installed. The code permits an alternative for engineered systems without insulation under the hydronic piping, provided that such insulation is shown to negatively affect the operation or efficiency of the heating system. To further improve energy efficiency, the new language also requires a thermal break of insulating material between the heated slab and the building foundation.

CHANGE TYPE: Addition

CHANGE SUMMARY: The 2009 IRC recognizes two new polyethylene materials and their associated fittings for hydronic piping.

2009 CODE:

M2104.3 Raised Temperature Polyethylene (PE-RT) Plastic Tubing. Joints between raised temperature polyethylene tubing and fittings shall conform to Sections M2104.3.1 and M2104.3.2. Mechanical joints shall be installed in accordance with the manufacturer's instructions.

M2104 and Table M2101.1 continues

TABLE M2101.1 Hydronic Piping Materials

Material	Use Code[a]	Standard[b]	Joints	Notes
Raised temperature polyethylene (PE-RT)	1, 2, 3	ASTM F 2623	Copper crimp/insert fitting stainless steel clamp, insert fittings	
Polyethylene/aluminum/polyethylene (PE-AL-PE) pressure pipe	1, 2, 3	ASTM F 1282 CSA B 137.9	Mechanical, crimp/insert	

(Portions of table and footnotes not shown remain unchanged.)

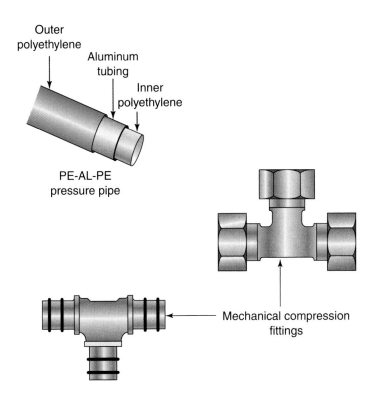

PE-AL-PE pressure pipe and fittings

*M2104 and Table M2101.1
continued*

M2104.3.1 Compression-Type Fittings. Where compression-type fittings include inserts and ferrules or O-rings, the fittings shall be installed without omitting the inserts and ferrules or O-rings.

M2104.3.2 PE-RT-to-Metal Connections. Solder joints in a metal pipe shall not occur within 18 inches (457 mm) of a transition from such metal pipe to PE-RT pipe.

M2104.4 Polyethylene/Aluminum/Polyethylene (PE-AL-PE) Pressure Pipe. Joints between polyethylene/aluminum/polyethylene pressure pipe and fittings shall conform to Sections M2104.3.1 and M2104.3.2. Mechanical joints shall be installed in accordance with the manufacturer's instructions.

M2104.4.1 Compression-Type Fittings. Where compression-type fittings include inserts and ferrules or O-rings, the fittings shall be installed without omitting the inserts and ferrules or O-rings.

M2104.4.2 PE-AL-PE to Metal Connections. Solder joints in a metal pipe shall not occur within 18 inches (457 mm) of a transition from such metal pipe to PE-AL-PE pipe.

CHANGE SIGNIFICANCE: Two additional polyethylene piping materials and their fittings are now approved for hydronic piping, typically hot water heating systems, for buildings regulated by the IRC. Raised temperature polyethylene (PE-RT) is a new generation of polyethylene resin tubing manufactured to ASTM F 2623 *Standard Specification for Polyethylene of Raised Temperature (PE-RT) SDR 9 Tubing*. PE-RT is manufactured for thermal stability, long-term pressure resistance, flexibility, resistance to corrosion and abrasion, and ease of installation. Polyethylene/aluminum/polyethylene (PE-AL-PE) pressure pipe is a composite pipe made of aluminum tube laminated to interior and exterior layers of polyethylene and manufactured to meet the testing requirements of ASTM F 1282 *Standard Specification for Polyethylene/ Aluminum/Polyethylene (PE-AL-PE) Composite Pressure Pipe*. PE-AL-PE pipe is identical to cross-linked polyethylene/aluminum/ cross-linked polyethylene (PEX/AL/PEX) composite pipe except for the physical properties of the polyethylene, which does not have the cross-linked molecular structuring of PEX. In the 2009 IRC, both PE-AL-PE and PEX/AL/PEX are approved for hot and cold potable water supply as well as hydronic piping for comfort heating. Similar to metal tubing, these aluminum core pipes bend and maintain their shape without restraint by attachment fittings.

Fuel gas as a special system is covered in Part 6, where issues such as fuel gas pipe design and installation, fuel gas piping materials, joints, and other such issues are addressed. The fuel gas provisions of the IRC and the text of its various sections are taken directly from the International Fuel Gas Code (IFGC) and reprinted directly into the IRC. In order to make the correlation and the coordination of the two codes easier, after each fuel gas section of the IRC the original section of the IFGC is shown in parentheses.

The fuel gas portion of the IRC contains its own specific definitions in Section G2403 in addition to the general definitions found in Chapter 2 of the IRC. Provisions, tables, and figures in other sections address the technical issues of fuel gas systems, such as appliance gas input, capacity of fuel gas pipes for system design, piping support, flow controls, gas vent systems, compressed-natural-gas dispensing systems, and other fuel gas issues. ■

G2408.2.1 and G2408.6

Appliance Installation in Garages

CHANGE TYPE: Addition

CHANGE SUMMARY: Elevation of the ignition source is not required for gas-fired appliances installed in an enclosed space that does not open into the garage. A new provision requires appliances to be installed and connected in a way that does not strain the gas piping connections.

2009 CODE: **G2408.2.1 (305.3.1) Installation in Residential Garages.**   In residential garages where appliances are installed in a separate, enclosed space having access only from outside of the garage, such appliances shall be permitted to be installed at floor level, provided that the required combustion air is taken from the exterior of the garage.

G2408.6 (305.12) Avoid Strain on Gas Piping.   Appliances shall be supported and connected to the piping so as not to exert undue strain on the connections.

CHANGE SIGNIFICANCE: Except for appliances that are specifically listed as flammable vapor ignition resistant, any appliance ignition source must be elevated at least 18 inches above the floor when installed in a garage. An ignition source is defined as a flame, spark or hot surface capable of igniting flammable vapors or fumes, and includes appliance burners, burner ignitors, and electrical switching devices. Because vapors from gasoline and other flammable liquids often found in garages are heavier than air and accumulate close to the floor, the lowest 18 inches in a garage is considered a hazardous location for ignition sources. The elevation requirements intend to prevent an explosion hazard. The new provision clarifies that an enclosed room that may be located in a garage but has no openings into the garage and does not pull combustion air from the garage does not need to meet the

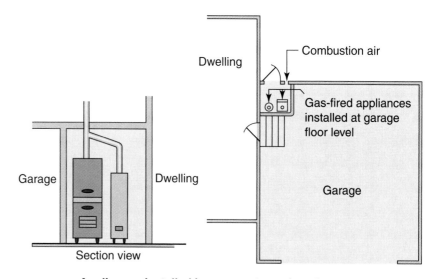

Appliances installed in a separate enclosed space

elevation requirements. Appliances installed in such enclosed spaces may be installed at the garage floor level.

The second part to this change prohibits placing any stress on the connections of the fuel gas piping system. The new language intends to reduce the possibility of damage to gas pipe fittings causing leaks or ignition of fuel gas.

G2415.4

Underground Penetrations Prohibited

CHANGE TYPE: Addition

CHANGE SUMMARY: Gas piping is no longer permitted to penetrate the foundation wall below ground.

2009 CODE: ~~**G2415.4 (404.4) Piping through Foundation Wall.** Underground piping, where installed below grade through the outer foundation or basement wall of a building, shall be encased in a protective pipe sleeve. The annular space between the gas piping and the sleeve shall be sealed.~~

<u>**G2415.4 (404.4) Underground penetrations prohibited.**</u> <u>Gas piping shall not penetrate building foundation walls at any point below grade. Gas piping shall enter and exit a building at a point above</u>

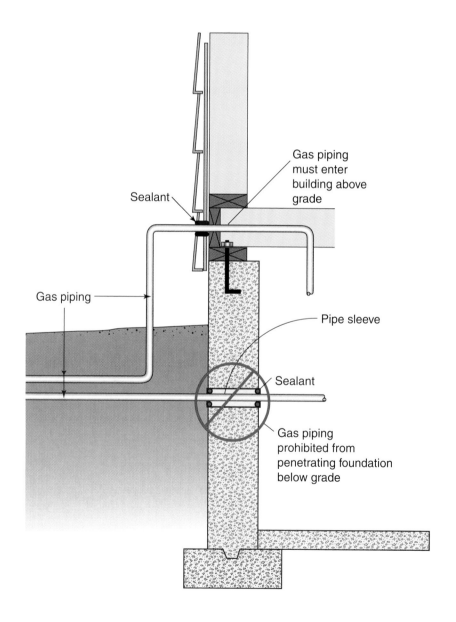

Sealant

Gas piping must enter building above grade

Gas piping

Pipe sleeve

Sealant

Gas piping prohibited from penetrating foundation below grade

Underground penetrations prohibited

<u>grade and the annular space between the pipe and the wall shall be sealed.</u>

CHANGE SIGNIFICANCE: It has been conventional practice for gas piping to enter a building from an outside underground location through the foundation into a crawl space, basement, or interior under-slab location. The 2006 IRC permitted such an installation provided the piping was in a sleeve and sealed against entry of water or gas. The 2009 IRC prohibits gas piping from penetrating a foundation wall below grade. The intent is that all gas piping entry and exit points of a building must be above ground. This would also preclude gas piping from passing beneath a foundation and entering the building through an interior slab on grade. The reason for this change is that underground gas leaks accumulate and follow the path of least resistance through the porous soil of the utility excavation to the building foundation. Further, underground entry points for gas piping provide a path for the leaking gas to enter the building and create an explosion hazard. The new measures intend to add a factor of safety for gas piping installations and require that all penetrations of the building envelope occur above ground.

G2415.6
Piping in Solid Floors

CHANGE TYPE: Modification

CHANGE SUMMARY: When both ends of a protective conduit installed in a slab terminate inside the building, the code now prohibits the ends from being sealed. For installations where one end of the conduit terminates outside the building, the interior end must be sealed and the exterior end must prevent entry of water and insects and be vented to the outdoors.

2009 CODE: G2415.6 (404.6) Piping in Solid Floors. Piping in solid floors shall be laid in channels in the floor and covered in a manner that will allow access to the piping with a minimum amount of damage to the building. Where such piping is subject to exposure to excessive moisture or corrosive substances, the piping shall be protected in an approved manner. As an alternative to installation in channels, the piping shall be installed in a conduit of Schedule 40 steel, wrought iron, PVC or ABS pipe ~~with tightly sealed ends and joints.~~ in accordance with Section G2415.6.1 or G2415.6.2. ~~Both of such conduit shall extend not less than 2 inches (51 mm) beyond the point where the pipe emerges from the floor The conduit shall be vented above grade to the outdoors and shall be installed so as to prevent the entry of water and insects.~~

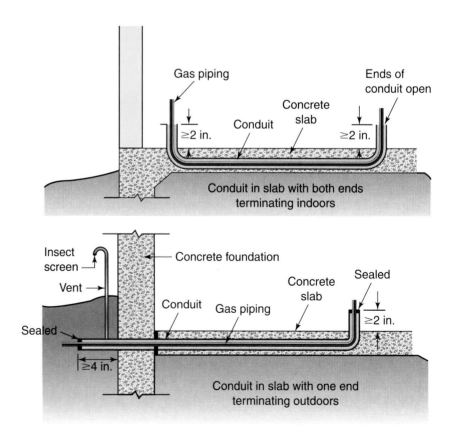

Gas piping in solid floors

G2415.6.1 (404.6.1) Conduit with One End Terminating Outdoors. The conduit shall extend into an occupiable portion of the building and, at the point where the conduit terminates in the building, the space between the conduit and the gas piping shall be sealed to prevent the possible entrance of any gas leakage. The conduit shall extend not less than 2 inches (51 mm) beyond the point where the pipe emerges from the floor. If the end sealing is capable of withstanding the full pressure of the gas pipe, the conduit shall be designed for the same pressure as the pipe. Such conduit shall extend not less than 4 inches (102 mm) outside of the building, shall be vented above grade to the outdoors and shall be installed so as prevent the entrance of water and insects.

G2415.6.2 (404.6.2) Conduit with Both Ends Terminating Indoors. Where the conduit originates and terminates within the same building, the conduit shall originate and terminate in an accessible portion of the building and shall not be sealed. The conduit shall extend not less than 2 inches (51 mm) beyond the point where the pipe emerges from the floor.

CHANGE SIGNIFICANCE: For gas piping installed in a conduit embedded in a concrete slab, the IRC now gives specific direction for treatment at the terminating ends of the conduit. The revisions recognize two installation scenarios for gas piping inside a conduit installed within a slab—one where both ends terminate inside the building and the other where one end of the conduit occurs inside and one outside. Previously, the code required the conduit end to be sealed where it extended out of the slab into the interior of the building. A vent to the outdoors prevented any buildup of leaking gas within the conduit. These provisions still apply when one end of the conduit terminates outside the building. For the other type of installation, with both ends terminating inside the building, the change supports a philosophy that it is better to leave the ends of the conduit unsealed so that any leaks in the gas piping can be detected within the building. In this case, the protective conduit is required to be open and located in an accessible space to enable quick detection by the building occupants should a gas leak occur. The open and unsealed conduit provides a similar level of gas leak detection that currently exists for a gas leak in any above-ground portion of the gas piping system.

G2415.12

Piping Underground Beneath Buildings

CHANGE TYPE: Modification

CHANGE SUMMARY: This change expands the installation requirements for gas piping encased by a conduit and installed beneath buildings. Where both ends of the protective conduit occur inside the building, the code now prohibits sealing of the conduit.

2009 CODE: G2415.12 (404.12) Piping Underground Beneath Buildings.   Piping installed underground beneath buildings is prohibited except where the piping is encased in a conduit of wrought iron, plastic pipe, steel pipe, <u>or other approved conduit material</u> designed to withstand the superimposed loads. ~~Such conduit shall extend into an occupiable portion of the building and, at the point where the conduit terminates in the building, the space between the conduit and the gas piping shall be sealed to prevent the possible entrance~~

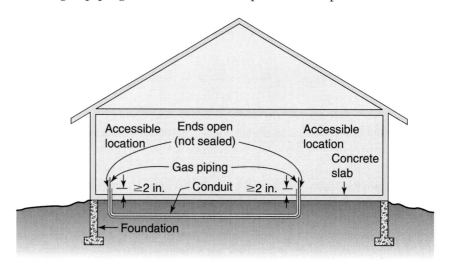

Conduit with both ends terminating indoors

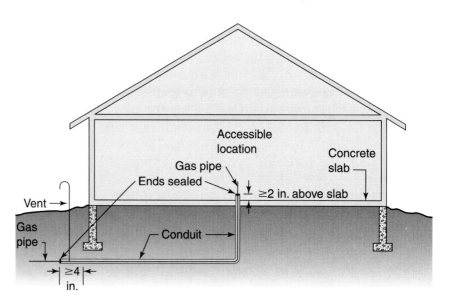

Conduit with one end terminating outdoors

Gas piping underground beneath buildings

~~of any gas leakage. If the end sealing is capable of withstanding the full pressure of the gas pipe, the conduit shall be designed for the same pressure as the pipe. Such conduit shall extend not less than 4 inches (102 mm) outside the building, shall be vented above grade to the outdoors, and shall be installed so as prevent the entrance of water and insects.~~ The conduit shall be protected from corrosion in accordance with Section G2415.9 <u>and shall be installed in accordance with Section G2415.12.1 or G2415.12.2</u>.

G2415.12.1 (404.12.1) Conduit with One End Terminating Outdoors.   <u>The conduit shall extend into an occupiable portion of the building and, at the point where the conduit terminates in the building, the space between the conduit and the gas piping shall be sealed to prevent the possible entrance of any gas leakage. The conduit shall extend not less than 2 inches (51 mm) beyond the point where the pipe emerges from the floor. Where the end sealing is capable of withstanding the full pressure of the gas pipe, the conduit shall be designed for the same pressure as the pipe. Such conduit shall extend not less than 4 inches (102 mm) outside the building, shall be vented above grade to the outdoors, and shall be installed so as to prevent the entrance of water and insects.</u>

G2415.12.2 (404.12.2) Conduit with Both Ends Terminating Indoors.   <u>Where the conduit originates and terminates within the same building, the conduit shall originate and terminate in an accessible portion of the building and shall not be sealed. The conduit shall extend not less than 2 inches (51 mm) beyond the point where the pipe emerges from the floor.</u>

CHANGE SIGNIFICANCE: The revision adds a second installation method for gas piping that is above ground indoors but must be extended beneath the slab to reach an appliance—for example, a cooking appliance located in a kitchen island. In this case, the protective conduit is required to be open (not sealed) and located in an accessible space to enable quick detection by the building occupants should a gas leak occur. The open and unsealed conduit provides a similar level of gas leak detection that currently exists for a gas leak in any above-ground portion of the gas piping system.

G2420.5

Appliance Shutoff Valve

CHANGE TYPE: Modification

CHANGE SUMMARY: Section G2420.5 has been reorganized to clarify the shutoff valve location requirements. When a manifold piping configuration is installed, a new provision permits the shutoff valve to be located at the manifold and not greater than 50 feet from the appliance. Because the two terms now have separate meanings, *appliance* replaces *equipment* in this section and in many other locations throughout the fuel gas provisions of the IRC.

2009 CODE: G2420.5 (409.5) ~~Equipment~~ Appliance Shutoff Valve. Each appliance shall be provided with a shutoff valve ~~separate from the appliance. The shutoff valve shall be located in the same room as the appliance, not further than 6 feet (1829 mm) from the appliance, and shall be installed upstream from the union, connector or quick disconnect device it serves. Such shutoff valves shall be provided with access.~~ in accordance with Section G2420.5.1, G2420.5.2 or G2420.5.3.

> **Exception:** ~~Shutoff valves for vented decorative appliances and decorative appliances for installation in vented fireplaces shall not be prohibited from being installed in an area remote from the appliance where such valves are provided with ready access. Such valves shall be permanently identified and shall serve no other equipment.~~

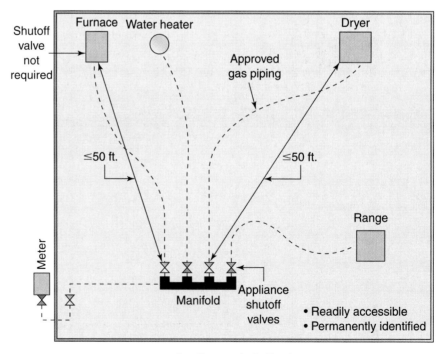

Appliance shutoff valves

~~**G2420.5.1 (409.5.1) Shutoff Valve in Fireplace.** Equipment shutoff valves located in the firebox of a fireplace shall be installed in accordance with the appliance manufacturer's instructions.~~

G2420.5.1 (409.5.1) Located Within Same Room. The shutoff valve shall be located in the same room as the appliance. The shutoff valve shall be within 6 feet (1829 mm) of the appliance and shall be installed upstream of the union, connector or quick disconnect device it serves. Such shutoff valves shall be provided with access. Appliance shutoff valves located in the firebox of a fireplace shall be installed in accordance with the appliance manufacturer's instructions.

G2420.5.2 (409.5.2) Vented Decorative Appliances and Room Heaters. Shutoff valves for vented decorative appliances, room heaters, and decorative appliances for installation in vented fireplaces shall be permitted to be installed in an area remote from the appliances where such valves are provided with ready access. Such valves shall be permanently identified and shall serve no other appliance. The piping from the shutoff valve to within 6 feet (1829 mm) of the appliance shall be designed, sized and installed in accordance with Sections G2412 through G2419.

G2420.5.3 (409.5.3) Located at Manifold. Where the appliance shutoff valve is installed at a manifold, such shutoff valve shall be located within 50 feet (15 240 mm) of the appliance served and shall be readily accessible and permanently identified. The piping from the manifold to within 6 feet (1829 mm) of the appliance shall be designed, sized and installed in accordance with Sections G2412 through G2419.

G2403 Definitions

> **Appliance ~~(equipment)~~.** Any apparatus or ~~equipment~~ device that uses gas as a fuel or raw material to produce light, heat, power, refrigeration or air conditioning.
>
> **Equipment.** ~~See "Appliance."~~ Apparatus and devices other than appliances.
>
> ~~**Fuel Gas Utilization Equipment.** See "Appliance."~~

CHANGE SIGNIFICANCE: The fuel gas provisions of the IRC require shutoff valves to permit the servicing and replacement of appliances without the need to shut down the entire gas supply system. These valves are not considered emergency shutoff valves. The reorganization of Section G2420.5 clarifies the shutoff valve location requirements and adds a provision for a valve located at the manifold when such a piping configuration is installed. While shutoff valves are typically required within 6 feet of the appliance, the valve located at the manifold may be as much as 50 feet from the appliance. The manifold is considered an appropriate location that does not impede the servicing or replacement of the appliance and does not require the entire gas piping system to be shut down. The phrase "separate from the ap-

G2420.5 continues

G2420.5 continued

pliance" has been deleted because it has mistakenly been interpreted as meaning that the shutoff valve cannot be located inside or under the housing of an appliance. The intent of this phrase was that the appliance's automatic valve cannot be used to meet the shutoff valve requirement. For some appliances, such as wall heaters and vented fireplaces, the shutoff valve may be installed inside or under the appliance.

Changes to the definitions of appliance and equipment clarify that they have different meanings and are not interchangeable, bringing greater precision to the code provisions. This is a global change to the fuel gas provisions of the IRC that substitutes the term *appliance* for the terms *equipment* and *gas utilization equipment* wherever they are currently used as a synonym for *appliance*. Since it matched the definition of appliance, the definition for *gas utilization equipment* has been deleted.

G2422.1.2.1
Maximum Length of Connectors

CHANGE TYPE: Modification

CHANGE SUMMARY: The maximum length of appliance connectors has increased from 3 feet to 6 feet.

2009 CODE: G2422.1.2.1 (411.1.3.1) Maximum Length. Connectors shall <u>not exceed</u> ~~have an overall length not to exceed 3 feet (914 mm)~~ 6 feet (1829 mm) in overall length ~~except for range and domestic clothes dryer connectors, which shall not exceed 6 feet (1829 mm) in overall length~~. Measurement shall be made along the centerline of the connector. Only one connector shall be used for each appliance.

> **Exception:** Rigid metallic piping used to connect an appliance to the piping system shall be permitted to have a total length greater than ~~3 feet (914 mm)~~ 6 feet (1829 mm), provided that the connecting pipe is sized as part of the piping system in accordance with Section G2413 and the location of the ~~equipment~~ <u>appliance</u> shutoff valve complies with Section G2420.5.

CHANGE SIGNIFICANCE: Previously, the limitation of 3 feet for most appliance connectors did not match the shutoff valve location requirements of G2420.5, which permits valves to be located 6 feet from the appliance, or the listing requirements of Section G2422.1. Connectors listed as conforming to the referenced standards ANSI Z21.24 and ANSI Z21.75 are permitted to be up to 6 feet in length according to

G2422.1.2.1 continues

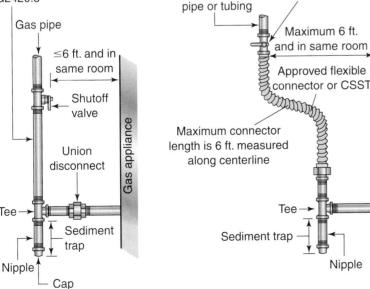

Rigid metallic piping is permitted to exceed 6 ft. if sized as part of gas piping system. Appliance shutoff must comply with Section G2420.5

Gas pipe

≤6 ft. and in same room

Shutoff valve

Union disconnect

Gas appliance

Tee →

Sediment trap

Nipple

Cap

Approved gas pipe or tubing   Shutoff valve

Maximum 6 ft. and in same room

Approved flexible connector or CSST

Maximum connector length is 6 ft. measured along centerline

Gas appliance

Tee →

Sediment trap

Nipple

Maximum length of appliance connectors

G2422.1.2.1 continued

the requirements of the standards. In addition, listed 6-foot connectors have an excellent safety record. Removing the differing length requirements based on the type of appliance clarifies the application of the code and eliminates overly restrictive language. This change also correlates the connector length provisions with IRC Sections G2420.5 and G2422.1, and provides consistency with ANSI Z223.1, *National Fuel Gas Code.*

G2439.5
Clothes Dryer Ducts

CHANGE TYPE: Modification

CHANGE SUMMARY: The reorganization of this section clarifies the code provisions for clothes dryer ducts. Except where determined by the manufacturer's installation instructions, the maximum prescribed length for gas dryer exhaust ducts has increased from 25 feet to 35 feet. Deductions from this maximum length for elbow fittings still apply but now appear in a table of equivalent lengths for fittings based on the radius and type of fitting. When dryer exhaust duct is concealed, the developed length must be identified with a label or tag located within 6 feet of the dryer. Provisions to protect dryer ducts from fastener penetration have been added and are consistent with vent and piping protection requirements.

2009 CODE: **G2439.5 (614.6) <u>Domestic</u> Clothes Dryer <u>Exhaust</u> Ducts.** Exhaust ducts for domestic clothes dryers shall ~~be constructed of metal and shall have a smooth interior finish.~~ <u>conform to the requirements of Sections G2439.5.1 through G2439.5.7.</u> ~~The exhaust duct shall be a minimum nominal size of 4 inches (102 mm) in diameter. The entire exhaust system shall be supported and secured in place. The male end of the duct at overlapped duct joints shall extend in the direction of airflow. Clothes dryer transition ducts used to connect the appliance to the exhaust duct system shall be metal and limited to a single length not to exceed 8 feet (2438 mm) in length and shall be listed and labeled for the application. Transition ducts shall not be concealed within construction.~~

G2439.5.1 (614.6.1) Material and Size. <u>Exhaust ducts shall have a smooth interior finish and shall be constructed of metal a minimum 0.016-inch (0.4 mm) thick. The exhaust duct size shall be 4 inches nominal in diameter.</u>

G2439.5 continues

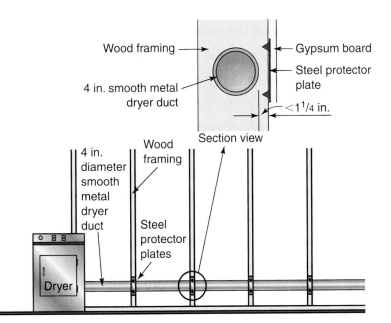

Clothes dryer duct protection against fastener penetration

G2439.5 continued

G2439.5.2 (614.6.2) Duct Installation. Exhaust ducts shall be supported at 4 foot intervals and secured in place. The insert end of the duct shall extend into the adjoining duct or fitting in the direction of airflow. Ducts shall not be joined with screws or similar fasteners that protrude into the inside of the duct.

G2439.5.3 (614.6.3) Protection Required. Protective shield plates shall be placed where nails or screws from finish or other work are likely to penetrate the clothes dryer exhaust duct. Shield plates shall be placed on the finished face of all framing members where there is less than 1¼ inches (32 mm) between the duct and the finished face of the framing member. Protective shield plates shall be constructed of steel, shall have a thickness of 0.062 inch (1.6 mm), and shall extend a minimum of 2 inches above sole plates and below top plates.

G2439.5.4 (614.6.4) Transition Ducts. Transition ducts used to connect the dryer to the exhaust duct system shall be a single length that is listed and labeled in accordance with UL 2158A. Transition ducts shall be a maximum of 8 feet in length and shall not be concealed within construction.

G2439.5.5 (614.6.5) Duct Length. The maximum allowable exhaust duct length shall be determined by one of the methods specified in Section G2439.5.5.1 or G2439.5.5.2.

G2439.5.1 (614.6.1) G2439.5.5.1 (614.6.5.1) ~~Maximum Length.~~ Specified Length. The maximum length of ~~a clothes dryer~~ the exhaust duct shall ~~be not exceed 25 feet (7620 mm)~~ 35 feet (10 668 mm) from the ~~dryer location~~ connection to the transition duct from the dryer to the outlet terminal. Where fittings are used, the maximum length of the exhaust duct shall be reduced in accordance with Table G2439.5.5.1. ~~The maximum length of a clothes dryer exhaust duct shall not exceed 25 feet (7620 mm) from to the outlet terminal. The maximum length of the duct shall be reduced 2½ feet (762 mm) for each 45 degree (0.79 rad) bend and 5 feet (1524 mm) for each 90 degree (1.6 rad) bend.~~

> **Exception:** ~~Where the make and model of the clothes dryer to be installed is known and the manufacturer's installation instructions for such dryer are provided to the code official, the maximum length of the exhaust duct, including any transition duct, shall be permitted to be in accordance with the dryer manufacturer's installation instructions.~~

G2439.5.5.2 (614.6.5.2) Manufacturer's Instructions. The maximum length of the exhaust duct shall be determined by the dryer manufacturer's installation instructions. The code official shall be provided with a copy of the installation instructions for the make and model of the dryer. Where the exhaust duct is to be concealed, the installation instructions shall be provided to the code official prior to the concealment inspection. In the absence of fitting equivalent length calculations from the clothes dryer manufacturer, Table G2439.5.5.1 shall be used.

Table G2439.5.5.1 (Table 614.6.5.1)　**Dryer Exhaust Duct Fitting Equivalent Length**

Dryer Exhaust Duct Fitting Type	Equivalent Length (feet)
4 inches radius mitered 45 degree elbow	2 feet 6 inches
4 inches radius mitered 90 degree elbow	5 feet
6 inches radius smooth 45 degree elbow	1 foot
6 inches radius smooth 90 degree elbow	1 foot 9 inches
8 inches radius smooth 45 degree elbow	1 foot
8 inches radius smooth 90 degree elbow	1 foot 7 inches
10 inches radius smooth 45 degree elbow	9 inches
10 inches radius smooth 90 degree elbow	1 foot 6 inches

G2439.5.6 (614.6.6) Length Identification. Where the exhaust duct is concealed within the building construction, the equivalent length of the exhaust duct shall be identified on a permanent label or tag. The label or tag shall be located within 6 feet of the exhaust duct connection.

G2439.5.2 (614.6.2) G2439.5.7 (614.6.7) Rough-In-Required. Exhaust Duct Required. Where a compartment or space for a clothes dryer is provided, an exhaust duct system shall be installed. Where the clothes dryer is not installed at the time of occupancy, the exhaust duct shall be capped at location of the future dryer.

CHANGE SIGNIFICANCE: Section G2439.5 has been completely reorganized and expanded to improve the technical accuracy and understanding of the exhaust duct provisions for gas clothes dryers. Changes are similar to those implemented for electric dryers in Section M1502 of the IRC mechanical provisions. Duct material, installation, length, and protection from damage are discussed separately in new subsections. Many of the provisions have not changed significantly, but there are some notable differences. Previously, the minimum nominal size for dryer ducts was 4 inches in diameter. Since air velocity drops dramatically when duct size is increased to 5 inches, the code now prescribes a nominal size of 4 inches and does not permit a larger duct diameter. The larger duct will not provide the minimum velocity to remove the moisture and any lint that gets into the exhaust duct. In addition, the IRC now specifies spacing not greater than 4 feet for duct supports. This is a typical spacing for supporting a 4-inch duct with insert joints and replaces the previous performance language requiring only that the exhaust system be supported.

There are also changes to the duct length requirements. By moving the provisions into separate sections, the code clarifies that there are two distinct paths for compliance—the prescriptive length that applies to any dryer or a length determined in accordance with the specific dryer manufacturer's installation instructions. When using the prescriptive requirements, typically because the make and model of the dryer are not known, the IRC now permits a maximum duct length of 35 feet from the connector to the duct termination. This increase from the previous 25 feet recognizes the improved efficiency of mod-

G2439.5 continues

G2439.5 continued

ern dryers and the fact that manufacturers' listings for most models far exceed the 35-foot restriction.

For installations following the prescriptive requirements, the code has simplified the length reductions for elbow fittings. The equivalent lengths for fittings have been placed in a table and recognize the reduced friction and improved air flow for larger-radius fittings. For example, a typical 4-inch radius 90-degree elbow deducts 5 feet from the allowable length of the dryer duct, where a 10-inch radius 90-degree elbow deducts only 1.5 feet. Equivalencies for the newer fittings were determined based on an analysis using the ASHRAE and SMACNA fitting tables.

Where dryer exhaust duct is concealed, the code now requires a permanent label or tag identifying the developed length of the exhaust duct. The identification must be located within 6 feet of the dryer. The distance is based on the maximum distance the gas shut-off valve can be located from a gas dryer. Since this has been used to determine close proximity for a gas valve, it is appropriate to use the same distance for close proximity for a label. This new requirement intends to alert homeowners installing replacement dryers to match the specifications for the make and model to the existing exhaust duct installation.

The 2009 IRC also adds provisions for protecting dryer ducts from penetration by fasteners. The new requirements are similar to protection requirements for piping and gas vents. In the case of dryer ducts, the concern of fastener penetration is related to the buildup of lint catching on the penetrating fasteners over time and increasing the fire hazard.

CHANGE TYPE: Addition

CHANGE SUMMARY: Clearance requirements for household cooking appliances have been added to the fuel gas provisions of the 2009 IRC. In general, kitchen wall cabinets must be positioned at least 30 inches above the surface of the range or cook top. Reduced clearances are permitted for the installation of listed appliances or fire-resistant insulating material below the wall cabinets and above the cooking appliance.

2009 CODE: G2447.5 (623.7) Vertical Clearance Above Cooking Top.   Household cooking appliances shall have a vertical clearance above the cooking top of not less than 30 inches (760 mm) to combustible material and metal cabinets. A minimum clearance of 24 inches (610 mm) is permitted where one of the following is installed:

1. The underside of the combustible material or metal cabinet above the cooking top is protected with not less than ¼-inch (6 mm)-thick insulating millboard covered with sheet metal not less than 0.0122 inch (0.3 mm) thick.

2. A metal ventilating hood constructed of sheet metal not less than 0.0122 inch (0.3 mm) thick is installed above the cooking top with a clearance of not less than ¼ inch (6 mm) between the hood and the underside of the combustible material or metal cabinet. The hood shall have a width not less than the width of the appliance and shall be centered over the appliance.

G2447.5 continues

G2447.5

Vertical Clearance Above Cooking Top

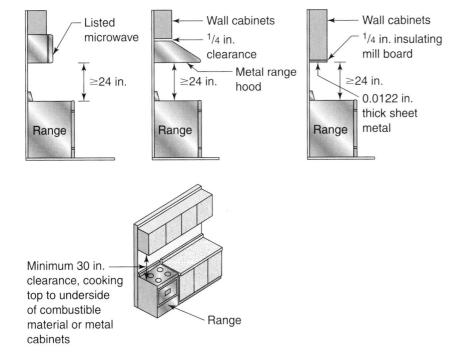

Vertical clearance above cooking top

G2447.5 continued

3. <u>A listed cooking appliance or microwave oven is installed over a listed cooking appliance and in compliance with the terms of the manufacturer's installation instructions for the upper appliance.</u>

CHANGE SIGNIFICANCE: New provisions for clearances above gas-fired residential cooking stoves generally match those of Section M1901.1. In both sections, a minimum clearance of 30 inches is required between the cooking surface and combustible materials above. Because metal cabinets readily conduct heat to the cabinet contents, Section G2447.5 also requires the 30-inch clearance below metal cabinets. These requirements in the mechanical and fuel gas provisions of the IRC intend to reduce the hazard of kitchen fires. Section M1901.1 permits reduced clearances in accordance with the listing and labeling of range hoods or appliances installed above the stove. Section G2447.5 is more specific in describing the conditions for reducing the clearance dimensions and permits a reduction to not less than 24 inches above the cooking surface. Certain microwaves and other cooking appliances are manufactured specifically for such an installation and are permitted to be installed with 24 inches of clearance when allowed by the manufacturer's installation instructions. The other two exceptions use a combination of metal and ¼-inch air space or ¼-inch insulating millboard to protect the underside of the cabinets. The new language is consistent with text in ANSI Z223.1, *National Fuel Gas Code (NFGC).*

PART 7

Plumbing

Chapters 25 through 33

Part 7 of the IRC contains provisions for plumbing systems. It includes a chapter on the specific and unique administrative issues of plumbing enforcement as well as the technical subjects for the overall design and installation of building plumbing systems. General plumbing issues such as the protection of piping systems from damage, piping support, and workmanship are covered in Chapter 26. The other chapters of Part 7 cover specific plumbing subjects: plumbing fixtures, water heaters, water supply and distribution, drainage, vents, and traps. ■

P2705.1

Installation of Fixtures

P2719.1

Floor Drains

P2902.6

Location of Backflow Preventers

P2904

Dwelling Fire Sprinkler Systems

P3005.2.6

Cleanout at the Base of Stacks

P3007

Sumps and Ejectors

P3108.1 AND P3108.2

Wet Venting

P2705.1

Installation of Fixtures

CHANGE TYPE: Modification

CHANGE SUMMARY: Bathroom fixture clearance dimensions have been revised to include lavatories and to provide consistency with the requirements of the *International Plumbing Code* (IPC).

2009 CODE: P2705.1 General. The installation of fixtures shall conform to the following:

1. **(through) 4.** (No change to text.)

5. ~~The centerline of water closets or bidets shall not be less than 15 inches (381 mm) from adjacent walls or partitions or not less than 15 inches (381 mm) from the centerline of a bidet to the outermost rim of an adjacent water closet. There shall be at least 21 inches (533 mm) clearance in front of the water closet, bidet or lavatory to any wall, fixture or door.~~

5. <u>Water closets, lavatories and bidets. A water closet, lavatory or bidet shall not be set closer than 15 inches (381 mm) from its center to any side wall, partition, or vanity or closer than 30 inches (762 mm) center-to-center between adjacent fixtures.</u>

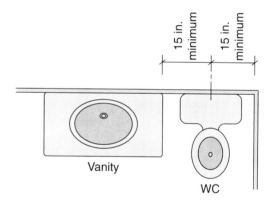

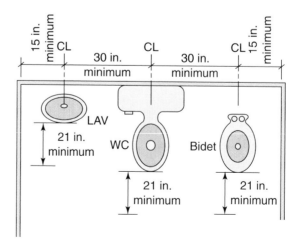

Installation of fixtures

> There shall be at least a 21-inch (533 mm) clearance in front of the water closet, lavatory, or bidet to any wall, fixture, or door.

6. (through) 8. (No change to text.)

CHANGE SIGNIFICANCE: The IRC plumbing provisions now provide clear direction on the spacing of water closets, bidets, and lavatories that is consistent with IPC Section 405.3.1. Previously, lavatory spacing was not included, and the measurement between a bidet and a water closet was subjective and contrary to the IPC. Now all adjacent fixtures require minimum 30-inch spacing measured centerline to centerline. Vanities have been added to the list of objects requiring a clearance of not less than 15 inches measured from the centerline of an adjacent fixture. Vanities are not fixtures, and this change clarifies that the centerline-to-centerline measurement for adjacent fixtures applies to wall-hung or pedestal lavatories, not sinks set in or on vanity cabinets.

P2719.1

Floor Drains

CHANGE TYPE: Clarification

CHANGE SUMMARY: This change clarifies that floor drains are not permitted beneath fixed appliances such as furnaces and water heaters or in inaccessible areas behind such appliances.

2009 CODE: P2719.1 Floor Drains. Floor drains shall have waste outlets not less than 2 inches (51 mm) in diameter and shall be provided with a removable strainer. The floor drain shall be constructed so that the drain is capable of being cleaned. Access shall be provided to the drain inlet. Floor drains shall not be located under or have their access restricted by permanently installed appliances.

CHANGE SIGNIFICANCE: Floor drains are not specifically required by the IRC. However, they are often installed as receptors for heating and cooling condensate or the discharge of the pressure and temperature relief valves of water heaters. It follows that floor drains are typically installed in utility rooms containing furnaces and water heaters. This change to the IRC clarifies that floor drains require access for maintenance and cleaning. The code now specifically prohibits floor drains from being installed beneath furnaces, water heaters, or other permanently installed appliances or in a location made inaccessible by any such appliance.

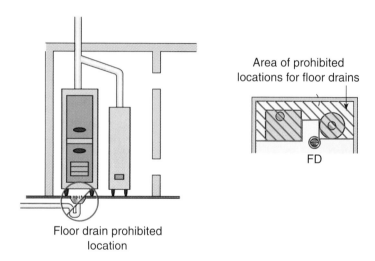

Floor drain prohibited
location

Area of prohibited
locations for floor drains

FD

Access to floor drains required

P2902.6
Location of Backflow Preventers

CHANGE TYPE: Clarification

CHANGE SUMMARY: This change adds guidelines for the installation, location, and protection of backflow preventers in accordance with the manufacturer's instructions and referenced standards.

2009 CODE: **P2902.6** ~~Access~~ **Location of Backflow Preventers.** ~~Backflow prevention devices shall be accessible for inspection and servicing.~~ Access shall be provided to backflow preventers as specified by the manufacturer's installation instructions.

P2902.6.1 Outdoor Enclosures for Backflow Prevention Devices. Outdoor enclosures for backflow prevention devices shall comply with ASSE 1060.

P2902.6.2 Protection of Backflow Preventers. Backflow preventers shall not be located in areas subject to freezing except where they can be removed by means of unions or are protected by heat, insulation, or both.

P2902.6.3 Relief Port Piping. The termination of the piping from the relief port or air gap fitting of the backflow preventer shall discharge to an approved indirect waste receptor or to the outdoors, where it will not cause damage or create a nuisance.

CHANGE SIGNIFICANCE: The revised and expanded backflow preventer provisions in Section P2902.6 correlate, in part, to the IPC and bring other language into the code that is usually contained in the manufacturer's installation instructions. The change is intended as a clarification of location and protection requirements that improve the understanding and usability of the code. Because of different in-

P2902.6 continues

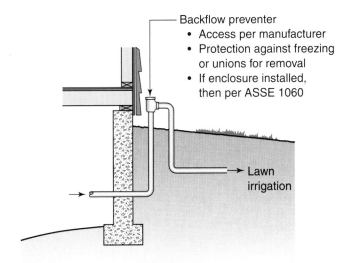

Backflow preventer
- Access per manufacturer
- Protection against freezing or unions for removal
- If enclosure installed, then per ASSE 1060

Lawn irrigation

Location of backflow preventers

P2902.6 continued stallation and service requirements for the various makes and types of backflow preventers, provisions for access are determined by the manufacturer.

The new referenced standard, ASSE 1060 *Performance Requirements for Outdoor Enclosures for Backflow Prevention Assemblies*, establishes criteria for protecting backflow preventers against physical damage and freezing while maintaining means for drainage and easy access for testing and maintenance. Many types of protection enclosures are manufactured that conform to ASSE 1060, including models with insulation and provisions for heating in below-freezing temperatures. Backflow preventers used in residential construction are typically installed in conditioned areas. For those subject to freezing temperatures, the code now clearly requires a means for easy removal during the winter or protection against freezing by the use of insulated or heated enclosures. Note that Sections P2902.6.1 and P2906.2 do not require enclosures for backflow preventers installed outdoors, but when enclosures are installed, they must conform to ASSE 1060. Consistent with provisions for condensate and temperature and pressure relief valve discharge, relief port piping from backflow preventers must discharge to an approved location that does not cause damage or create a nuisance.

CHANGE TYPE: Addition

CHANGE SUMMARY: The new IRC Section P2904 provides a simple, prescriptive approach for the design of dwelling fire sprinkler systems without referencing NFPA 13D.

P2904
Dwelling Fire Sprinkler Systems

2009 CODE:

P2904.1 General. Where installed, residential fire sprinkler systems, or portions thereof, shall be in accordance with NFPA 13D or Section P2904, which shall be considered equivalent to NFPA 13D. Section P2904 shall apply to stand-alone and multipurpose wet-pipe sprinkler systems that do not include the use of antifreeze. A multipurpose fire sprinkler system shall provide domestic water to both fire sprinklers and plumbing fixtures. A stand-alone sprinkler system shall be separate and independent from the water distribution system. A backflow preventer shall not be required to separate a stand-alone sprinkler system from the water distribution system.

P2904 continues

Photo courtesy of Uponor Inc.

P2904 continued

P2904.1.1 Required Sprinkler Locations. Sprinklers shall be installed to protect all areas of a dwelling unit.

Exceptions:

1. Attics, crawl spaces, and normally unoccupied concealed spaces that do not contain fuel-fired appliances do not require sprinklers. In attics, crawl spaces, and normally unoccupied concealed spaces that contain fuel-fired equipment, a sprinkler shall be provided above the equipment; however, sprinklers shall not be required in the remainder of the space.

2. Clothes closets, linen closets, and pantries not exceeding 24 square feet in area, with the smallest dimension not greater than 3 feet and having wall and ceiling surfaces of gypsum board

3. Bathrooms not greater than 55 square feet in area

4. Garages; carports; exterior porches; unheated entry areas, such as mud rooms, that are adjacent to an exterior door; and similar areas

The entire code change text for Section P2904 is too extensive to be included here. Refer to Code Change RP3-07/08 in the *2009 IRC Code Changes Resource Collection* for the complete text and history of the code change.

CHANGE SIGNIFICANCE: New IRC Section P2904 introduces a simple, prescriptive approach to the design of dwelling fire sprinkler systems. An automatic sprinkler system conforming to IRC Section P2904 or NFPA 13D, *Installation of Sprinkler Systems in One- and Two-Family Dwellings and Manufactured Houses*, is now required fire protection in all new one- and two-family dwellings and townhouses. International Residential Code (IRC) Section 313.1 requires the installation of fire sprinkler systems in one- and two-family dwellings after January 1, 2011. In townhouses, a residential automatic sprinkler system is required upon adoption of the 2009 IRC. A dwelling automatic sprinkler system aids in the detection and control of fires in residential occupancies and is expected to prevent total fire involvement (flashover) in the room of fire origin. A properly installed and maintained automatic sprinkler system complying with IRC Section P2904 or NFPA 13D improves the likelihood of occupants escaping or being evacuated.

A dwelling fire sprinkler system requires less water when compared with NFPA 13 and 13R systems. Section P2904.5.2 requires a minimum water discharge duration of 10 minutes, compared with 30 minutes for an NFPA 13R system and even higher duration values in NFPA 13. A water supply that can supply the minimum sprinkler discharge density for an NFPA 13D automatic sprinkler system such as a connection to a domestic water supply, a water well, an elevated storage tank, a pressure tank constructed in accordance with the American Society of Mechanical Engineers standards that is connected to a reliable source of air pressure, or a stored water source with an automatically operated pump is permitted. Any combination of water supply systems is allowed to meet the required dwelling fire sprinkler system capacity. Section P2904.5.2 allows the supply duration to be reduced to seven minutes in one-story dwellings with an area of 2000 square feet or less.

Section P2904.1 requires the use of new sprinklers listed for residential applications. These sprinklers have been investigated and listed for use inside of dwelling units. They are equipped with a thermal element using either a fusible link or frangible bulb filled with conductive liquid that is designed to operate approximately five times faster when compared with a standard spray sprinkler required by NFPA 13 installed in the same setting. The sprinkler discharge pattern is designed to wet the walls and floors of the space and apply water onto any furnishings. The amount of water that the sprinkler is required to discharge is based on the geometry of the room or area being protected and the sprinkler's listing. At a minimum, NFPA 13D requires a discharge density of 0.05 gallon per minute (gpm) per square foot to the design sprinklers. In a hypothetical rectangular 120 square foot bedroom, this discharge density equals a flow rate of 6 gpm.

Compared with an NFPA 13 automatic sprinkler system, a dwelling fire sprinkler system does not require automatic sprinkler protection throughout a one- and two-family dwelling or townhouse. Sprinklers are not required in areas of one- and two-family dwellings and townhomes that have been statistically shown through fire incident loss data to not significantly contribute to injuries or death. Section P2904.1.1 omits sprinklers in closets and pantries with an area less than 24 square feet constructed of gypsum board walls and ceilings, and bathrooms with an area of 55 square feet or less. Sprinklers are not required in open attached porches, garages, attics, crawl spaces, and concealed spaces not intended or used for living purposes. In attics housing fuel-fired appliances, Section 2904.1.1 requires a single sprinkler above the equipment but does not require that the sprinkler protection be provided throughout the entire attic area. A dwelling fire sprinkler system does not require a fire department connection.

Automatic sprinkler systems are designed to supply a minimum volume of water over a given area served by two or more sprinklers. Pipes and fittings are sized based on the available water supply, which has a known flow rate and discharge pressure. Adjustments are made in pipe diameters and arrangements based on the location of the sprinklers in relation to the water supply and the length of pipe and fittings that the water must travel from the water supply to the most remote sprinklers. These adjustments are necessary to accommodate pressure changes due to elevation changes and friction loss of the water moving through the pipe. Fire protection designers and engineers commonly utilize commercially available software to perform these calculations based on mathematical equations that take into account the internal diameter and surface characteristics of the pipe and the minimum flow rate required by the sprinkler.

The prescriptive language of these new provisions in Section P2904 will allow a contractor or home builder to design and install a residential sprinkler system without referencing NFPA 13D. The requirements in Section P2904 are consistent with the requirements in

P2904 continues

P2904 continued

NFPA 13D, but have been simplified. Design professionals still have the option of using NFPA 13D, which allows for engineered design options and other piping configurations.

The approach of using prescriptive tabular schedules in the IRC but still permitting engineered design alternatives based on recognized standards is currently allowed for other building features. For example, consider the IRC's approach to structural design. In the case of floor systems, the IRC provides prescriptive span tables as a simple basis for conventional construction, but Section R301 gives the homebuilder an option to use an engineered design based on the IBC and ASCE 7, if desired. A fundamental assumption of Section P2904 is that piping will comply with all of the requirements applicable to a residential plumbing system established by IRC Chapters 25–29. For this reason, there is no need to address special subjects—for example, freeze protection—in P2904 because all residential plumbing is required to be protected from freezing, and installers of potable water systems will be familiar with local requirements for freeze protection. Another fundamental requirement is contained in Section P2904.6.2.2, which requires the use of the manufacturer's instructions for sprinklers and sprinkler pipe and that the instructions will include all of the basic requirements necessary to design and install these components.

The prescriptive pipe sizing method contained in Section P2904.6.2.1 is used to calculate the available pressure after deducting for various pressure losses resulting from friction loss in piping, the selected pipe and fittings, the diameter and length of the pipe, the pressure loss through the water service pipe and meter, and pressure losses due to elevation changes. Equation 29-1 also requires that the maximum pressure required for the sprinkler, which is specified by the manufacturer, be included in the calculation for the available pressure.

$$P_t = P_{sup} - PL_{svc} - PL_m - PL_d - PL_e - P_{sp}$$ **(Equation 29-1)**

where:

P_t = pressure used in applying Tables P2904.6.2(4) through P2904.6.2(9)

P_{sup} = pressure available from the water supply source

PL_{svc} = pressure loss in the water-service pipe

PL_m = pressure loss in the water meter

PL_d = pressure loss from devices other than the water meter

PL_e = pressure loss associated with changes in elevation

P_{sp} = maximum pressure required by a sprinkler

Inspection requirements for dwelling fire sprinkler systems are set forth in Section P2904.8. The IRC has pre-concealment inspection criteria in Section P2904.8.1 and final inspection criteria in Section P2904.8.2. The pre-concealment (a.k.a. rough-in) inspection includes evaluating required locations for sprinklers, the locations that may obstruct the sprinkler discharge pattern, whether the selected sprinklers have the correct temperature rating and are adequately separated from external heat sources, the correct pipe material, whether diameters and lengths are installed and are properly anchored, and witnessing a pressure test of the system. The final inspection includes a survey of sprinklers to ensure that they are not painted or damaged and that devices such as water softeners or water filters that can impair the water flow are not installed. In dwellings equipped with a pump, a test to ensure that the pump starts upon a water demand is required. The final inspection will also confirm the installation of a sign required by Section P2904.7 at the main water shutoff valve and that an owner's manual is made available to the owner.

P3005.2.6

Cleanout at the Base of Stacks

CHANGE TYPE: Modification

CHANGE SUMMARY: A cleanout is now required at the base of each sanitary drainage stack. Alternative locations near the stack or outside the building are no longer permitted.

2009 CODE: P3005.2.6 Base of Stacks. ~~Accessible cleanouts~~ A cleanout shall be provided ~~near~~ at the base of each ~~vertical~~ waste or soil stack. ~~Alternatively, such cleanouts shall be installed outside the building within 3 feet (914 mm) of the building wall.~~

CHANGE SIGNIFICANCE: The language describing an alternative location for cleanouts outside the building was considered ambiguous and has been deleted. While the cleanout was required to be within 3 feet of the building wall, there was no indication as to the permitted distance from the stack location or even if the cleanout needed to be located adjacent to the same wall as the stack. In addition, the stack was not required to be adjacent to the exterior wall in order to take advantage of the alternative and could conceivably be located in the middle of the building. Similarly, the location described as *near* the base of the stack seemed to indicate that the cleanout could be located some undefined horizontal distance from the stack. The 2009 IRC clearly requires a cleanout at the base of each sanitary drainage stack as intended by the code and does not permit alternative locations.

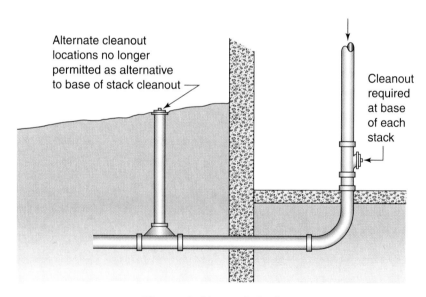

Cleanout at base of stacks

CHANGE TYPE: Modification

CHANGE SUMMARY: The provisions for sumps and ejectors in Section P3007 have been replaced to match the provisions in Section 712 of the IPC and to provide more comprehensive coverage of the requirements.

2009 CODE: **P3007.1 Building Subdrains.** Building subdrains that cannot be discharged to the sewer by gravity flow shall be discharged into a tightly covered and vented sump from which the liquid shall be lifted and discharged into the building gravity drainage system by automatic pumping equipment or other approved method. In other than existing structures, the sump shall not receive drainage from any piping within the building capable of being discharged by gravity to the building sewer.

P3007.2 Valves Required. A check valve and a full open valve located on the discharge side of the check valve shall be installed in the pump or ejector discharge piping between the pump or ejector and the gravity drainage system. Access shall be provided to such valves. Such valves shall be located above the sump cover required by Section P3007.3.2 or, where the discharge pipe from the ejector is below grade, the valves shall be accessibly located outside the sump below grade in an access pit with a removable access cover.

P3007.3.2 Sump Pit. The sump pit shall be not less than 18 inches (457 mm) in diameter and 24 inches (610 mm) deep, unless otherwise approved. The pit shall be accessible and located such that all drainage flows into the pit by gravity. The sump pit shall be constructed of tile, concrete, steel, plastic or other approved materials. The pit bot-

P3007 continues

P3007
Sumps and Ejectors

Sewage ejector installation

P3007 continued

TABLE P3007.6 **Minimum Capacity of Sewage Pump or Sewage Ejector**

Diameter of the Discharge Pipe (inches)	Capacity of Pump or Ejector (gpm)
2	21
2½	30
3	46

tom shall be solid and provide permanent support for the pump. The sump pit shall be fitted with a gas-tight removable cover adequate to support anticipated loads in the area of use. The sump pit shall be vented in accordance with Chapter 31.

Because this code change replaced all of Section P3007, the entire code change text is too extensive to be included here. Refer to Code Change RP22-06/07 in the *2009 IRC Code Changes Resource Collection* for the complete text and history of the code change.

CHANGE SIGNIFICANCE: This change imports the language from Section 712 of the IPC to completely replace the sump and ejector provisions of Section P3007. While the organization of the material has changed and the coverage is expanded, many of the previous IRC provisions remain in place. New provisions detail the size and construction of the sump pit. The minimum size is 18 inches in diameter and 24 inches deep. Approved materials include tile, concrete, steel, and plastic. Where the 2006 IRC required the sump to be tightly covered, the new language specifies a gas-tight removable cover. In addition, the code now prescribes controls to maintain the effluent level at least 2 inches below the gravity drain inlet.

Rather than specifying the minimum discharge velocity of the ejector pump, the new provisions include a performance statement that the sewage pump or ejector must have the capacity and head suitable for the application. To ensure adequate velocity, the code sets minimum pipe sizing and gallons per minute (gpm) ratings as listed in a new Table P3007.6.

Additional coverage is also provided for the discharge connection to the gravity drainage system. Unless connecting directly to the building sewer, the discharge connection to the building drain must be not less than 10 feet away from the base of any soil stack, waste stack, or fixture drain. Connection to either a vertical or horizontal portion of the drainage system must be accomplished through a wye fitting.

P3108.1 and P3108.2
Wet Venting

CHANGE TYPE: Modification

CHANGE SUMMARY: This change clarifies that each fixture drain must connect individually to the horizontal wet vent and is now consistent with the vertical wet venting provisions. A water closet is now permitted to be located upstream of the dry vent connection to the horizontal wet vent.

2009 CODE: P3108.1 Horizontal Wet Vent Permitted. Any combination of fixtures within two bathroom groups located on the same floor level are permitted to be vented by a horizontal wet vent. The wet vent shall be considered the vent for the fixtures and shall extend from the connection of the dry vent along the direction of the flow in the drain pipe to the most downstream fixture drain connection. Each fixture drain shall connect horizontally to the horizontal branch being wet vented or shall have a dry vent. <u>Each wet-vented fixture drain shall connect independently to the horizontal wet vent.</u> Only the fixtures within the bathroom groups shall connect to the wet-vented horizontal branch drain. Any additional fixtures shall discharge downstream of the horizontal wet vent.

P3108.2 Dry Vent Connection. <u>The required dry-vent connection for wet-vented systems shall comply with Sections P3108.2.1 and P3108.2.2</u>

~~**P3108.2**~~ **P3108.2.1** ~~**Vent Connections.**~~ **Horizontal Wet Vent.** The dry vent connection ~~to the wet vent~~ <u>for a horizontal wet vent system</u> shall be an individual vent or common vent ~~to the lavatory, bidet, shower or bathtub~~ <u>for any bathroom group fixture, except an emergency floor drain.</u> ~~In vertical wet vent systems, the most upstream fixture drain connection shall be a dry-vented fixture drain connection.~~

P3108.1 and P3108.2 continues

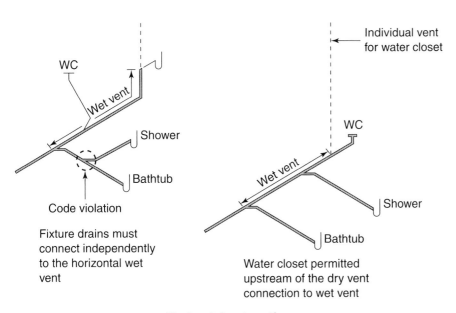

Horizontal wet venting

WC

Wet vent

Shower

Bathtub

Code violation

Fixture drains must connect independently to the horizontal wet vent

Individual vent for water closet

WC

Wet vent

Shower

Bathtub

Water closet permitted upstream of the dry vent connection to wet vent

P3108.1 and P3108.2 continued

~~In horizontal wet-vent systems,~~ Where the dry vent connects to a water closet fixture drain, the drain shall connect horizontally to the horizontal wet vent system. ~~n~~ Not more than one wet-vented fixture drain shall discharge upstream of the dry-vented fixture drain connection.

P3108.2.2 Vertical Wet Vent. The dry vent connection for a vertical wet vent system shall be an individual vent or common vent for the most upstream fixture drain.

CHANGE SIGNIFICANCE: While Section P3108.4 has always required each wet-vented fixture to connect independently to the vertical wet vent, the code has not included similar language in the horizontal wet vent section. A basic principle of wet venting is that each fixture connects directly to the wet vent. The new language clarifies the intent of the code and specifically prohibits two or more fixture drains joining before they connect to the wet vent. Sections P3108.1 and P3108.4 are now consistent.

The dry vent connection requirements have been broken into two sections to clarify the applications for horizontal and vertical wet vent systems. In horizontal wet vent systems, the individual vents of water closets are now included in the list of bathroom group fixtures that can serve as the required dry vent. This change recognizes that water closets are not restricted in locations upstream of the dry vent connection for circuit venting installations. Circuit venting is another form of wet venting that allows as many as eight fixtures on a branch but does not limit the fixtures to those contained within two bathroom groups. Similar to circuit venting principles, the water closet fixture drain must connect horizontally to the branch drain and horizontal wet vent system where the water closet is upstream of the dry vent connection. Vertical drains of water closets are not permitted as vents.

PART 8

Electrical

Chapters 34 through 43

The electrical portions of the IRC are contained within Chapter 34 through 43 and cover various electrical system design and installation issues, as well as definitions specific to electrical systems. It is intended that the electrical provisions contained within the IRC have no conflicts with the National Electrical Code (NEC), published by the National Fire Protection Association (NFPA); as such, the IRC electrical portions are taken (by permission of the NFPA) from the NEC. Appendix Q of the IRC cross-references the IRC electrical provisions to those of the NEC. The electrical portions of the IRC are organized in a similar fashion to the mechanical and plumbing portions, with chapters on general requirements, electrical services, wiring methods, power distribution, lighting fixtures, and other related issues. ■

E3607.3
Grounding for Buildings Supplied by Feeders

E3608.1
Grounding Electrodes

E3609.3
Intersystem Bonding Termination

E3705.7
Location of Overcurrent Devices

E3901
Required Receptacle Outlets

E3901.7
Outdoor Outlets

E3902.2 AND E3902.5
Ground-Fault Circuit-Interrupter (GFCI) Protection

E3902.11
Arc-Fault Protection

E4002.14
Tamper-Resistant Receptacles

E4003.12
Luminaires in Clothes Closets

E4203.3
Disconnecting Means for Pools, Spas, and Hot Tubs

E4206.5.1
Servicing of Wet-Niche Luminaires

E4209.1
Hydromassage Bathtubs

E3607.3

Grounding for Buildings Supplied by Feeders

CHANGE TYPE: Modification

CHANGE SUMMARY: For other than existing wiring systems, the code no longer permits feeders or branch circuits without an equipment grounding conductor to serve separate buildings.

2009 CODE: E3607.3 Buildings or Structures Supplied by Feeder(s) or Branch Circuit(s). Buildings or structures supplied by feeder(s) or branch circuit(s) shall have a grounding electrode or grounding electrode system installed in accordance with Section E3608. The grounding electrode conductor(s) shall be connected in a manner specified in Section E3607.3.1 or, for existing premises wiring systems only, E3607.3.2. Where there is no existing grounding electrode, the grounding electrode(s) required in Section E3608 shall be installed.

Exception: A grounding electrode shall not be required where only one branch circuit, including a multiwire branch circuit, supplies the building or structure and the branch circuit includes an equipment grounding conductor for grounding the non-current-carrying parts of all equipment. For the purposes of this section, a multiwire branch circuit shall be considered as a single branch circuit.

E3607.3.1 Equipment Grounding Conductor. An equipment grounding conductor as described in Section E3908 shall be run with the supply conductors and connected to the building or structure disconnecting means and to the grounding electrode(s). The equipment

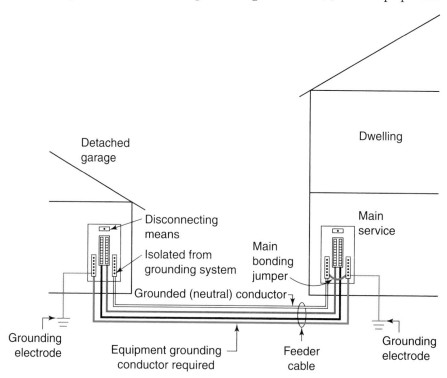

Grounding for separate building supplied by feeder

grounding conductor shall be used for grounding or bonding of equipment, structures or frames required to be grounded or bonded. The equipment grounding conductor shall be sized in accordance with Section E3908.12. Any installed grounded conductor shall not be connected to the equipment grounding conductor or to the grounding electrode(s).

E3607.3.2 Grounded Conductor, Existing Premises.   This section shall apply only to existing premises wiring systems. Where an equipment grounding conductor is not run with the supply conductors to the building or structure, ~~and~~ there are no continuous metallic paths bonded to the grounding system in both buildings or structures involved, and ground-fault protection of equipment has not been installed on ~~the common service~~ the supply side of the feeder(s), the grounded ~~circuit~~ conductor run with the supply ~~conductors~~ to the buildings or structure shall be connected to the building or structure disconnecting means and to the grounding electrode(s) and shall be used for grounding or bonding of equipment, structures, or frames required to be grounded ~~or bonded~~. Where used for grounding in accordance with this provision. ~~The size of~~ the grounded conductor shall not be smaller than the larger of:

1. That required by Section E3704.3.
2. That required by Section E3908.12.

CHANGE SIGNIFICANCE: Previously, the code permitted the common practice of installing a three-wire feeder to a detached accessory building without an equipment grounding conductor. In effect, the grounded conductor also served as the grounding conductor. Except for existing wiring systems, this practice is no longer allowed, and an equipment grounding conductor is always required when running a feeder or branch circuit to a separate building or structure. Typically this will result in the installation of a 120/240-volt four-wire feeder cable consisting of two ungrounded conductors, one grounded (neutral) conductor, and one equipment grounding conductor. Though a grounding electrode system is established at each building, the grounded conductor is bonded to the grounding system only at the main service and is isolated from the equipment grounding conductor and grounding system at the separate building. The IRC retains an exception for existing wiring systems that permits feeders without an equipment grounding conductor, provided the prescribed safeguards are in place.

E3608.1

Grounding Electrodes

CHANGE TYPE: Modification

CHANGE SUMMARY: For concrete-encased electrodes, the code clarifies the location of horizontal and vertical installations and stipulates that only one such electrode needs to be connected to the grounding electrode system. The provisions for rod and pipe electrodes have been revised to clarify the material, dimensions, and listing requirements. A new section recognizes other listed grounding electrodes.

2009 CODE: E3608.1.2 Concrete-Encased Electrode. An electrode encased by at least 2 inches (51 mm) of concrete, located <u>horizontally</u> ~~within and~~ near the bottom <u>or vertically and within that portion</u> of a concrete foundation or footing that is in direct contact with the earth, consisting of at least 20 feet (6096 mm) of one or more bare or zinc-galvanized or other electrically conductive coated steel reinforcing bars or rods of not less than ½ inch (12.7 mm) diameter, or consisting of at least 20 feet (6096 mm) of bare copper conductor not smaller than 4 AWG, shall be considered as a grounding electrode. Reinforcing bars shall be permitted to be bonded together by the usual steel tie wires or other effective means. <u>Where multiple concrete-encased electrodes are present at a building or structure, only one shall be required to be bonded into the grounding electrode system.</u>

E3608.1.3 Ground Rings. (No change to text.)

E3608.1.4 Rod and Pipe Electrodes. Rod and pipe electrodes not less than 8 feet (2438 mm) in length and consisting of the following materials shall be considered as a grounding electrode:

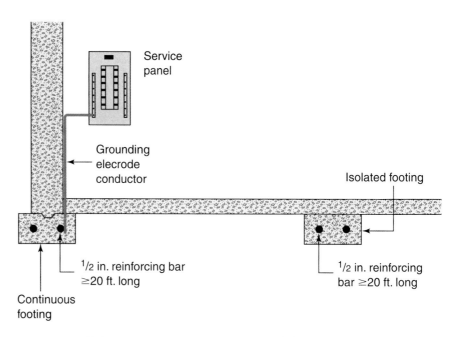

Only one concrete encased electrode is required to be bonded to the grounding system

1. <u>Grounding</u> electrodes of pipe or conduit shall not be smaller than trade size ¾ (metric designator 21) and, where of ~~iron or~~ steel, shall have the outer surface galvanized or otherwise metal-coated for corrosion protection.

2. <u>Grounding</u> electrodes of rods of <u>stainless steel and copper or zinc-coated</u> ~~iron or~~ steel shall be at least $\frac{5}{8}$ inch (15.9 mm) in diameter. Stainless steel rods less than $\frac{5}{8}$ inch (15.9 mm) in diameter, nonferrous rods or their equivalent shall be listed and shall be not less than ½ inch (12.7 mm) in diameter.

E3608.1.4.1 Installation. (No change to text.)

E3608.1.5 Plate Electrodes. (No change to text.)

<u>E3608.1.6 Other Electrodes.</u> <u>In addition to the grounding electrodes specified in Sections E3608.1.1 through E3608.1.5 other listed grounding electrodes shall be permitted.</u>

CHANGE SIGNIFICANCE: Concrete-encased grounding electrodes have proven very effective and do not require supplemental electrodes. This type of electrode consists of 20 feet of ½-inch reinforcing bar or 20 feet of 4 AWG bare copper wire encased in concrete in contact with the ground and covered with at least 2 inches of concrete. If reinforcing is present in a concrete footing, the code requires it to be used as part of the grounding electrode system. However, the 2009 IRC clarifies that separate concrete-encased electrodes do not have to be bonded together and that only one such electrode must be connected to the grounding electrode system. New language also clarifies that these electrodes are permitted in either a horizontal or vertical position. Vertical electrodes must be in a section of concrete foundation wall or pier that is in direct contact with the ground. Horizontal electrodes are installed near the bottom, typically in the footing.

Revisions to the rod and pipe electrode provisions have clarified the material and dimension requirements. The code no longer recognizes iron rod or pipe for use as a grounding electrode. Rods must be stainless steel or steel that is clad with copper or zinc. Previously, the text did not specifically address listed copper or zinc-coated steel rod electrodes, only those made of stainless steel or equivalent, but the intent was that listed *nonferrous rods* referred to copper and zinc coatings. The new language clearly requires the three types of steel rods to be at least $\frac{5}{8}$ inch in diameter or to be listed and not less than ½ inch in diameter.

By adding Section E3608.1.6, the code specifically recognizes new technology in producing other types of listed grounding electrodes. These are typically chemical or electrolytic type grounding electrodes that consist of copper rods or pipes filled with chemicals or salts that lower the resistance of the surrounding soil and enhance the grounding capabilities. Some installations rely on special backfill materials to further improve performance of these systems.

E3609.3

Intersystem Bonding Termination

CHANGE TYPE: Modification

CHANGE SUMMARY: Bonding terminations for communications, satellite, and cable television grounding conductors are now required in one of three prescribed and accessible locations.

2009 CODE: E3609.3 Bonding to for Other Systems. An intersystem bonding termination for connecting intersystem bonding and grounding conductors required for other systems shall be provided external to enclosures at the service equipment and at the disconnecting means for any additional buildings or structures. The intersystem bonding termination shall be accessible for connection and inspection. The intersystem bonding termination shall have the capacity for connection of not less than three intersystem bonding conductors. The intersystem bonding termination device shall not interfere with the opening of a service or metering equipment enclosure. The intersystem bonding termination shall be one of the following:

1. A set of terminals securely mounted to the meter enclosure and electrically connected to the meter enclosure. The terminals shall be listed as grounding and bonding equipment.

2. A bonding bar near the service equipment enclosure, meter enclosure, or raceway for service conductors. The bonding bar shall be connected with a minimum 6 AWG copper conductor to an equipment grounding conductor(s) in the service equipment enclosure, to a meter enclosure, or to an exposed non-flexible metallic raceway.

3. A bonding bar near the grounding electrode conductor. The bonding bar shall be connected to the grounding electrode conductor with a minimum 6 AWG copper conductor.

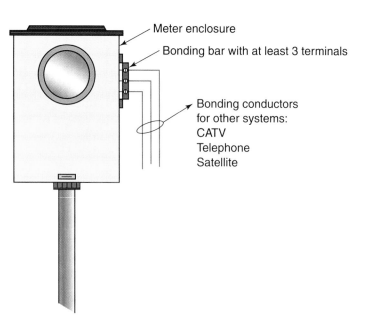

Intersystem bonding termination

~~An accessible means external to enclosures for connecting intersystem bonding and grounding electrode conductors shall be provided at the service equipment and at the disconnecting means for any additional buildings or structures by at least one of the following means:~~

1. ~~Exposed nonflexible metallic raceways~~
2. ~~An exposed grounding electrode conductor~~
3. ~~Approved means for the external connection of a copper or other corrosion-resistant bonding or grounding conductor to the grounded raceway or equipment~~

CHANGE SIGNIFICANCE: For bonding of other systems—typically telephone, satellite, and cable television systems—to the building grounding system, the installer has relied on locating an exposed section of metal raceway, grounding electrode conductor, or bonded metal water pipe. Such installations have resulted in these other systems being bonded to the dwelling electrical system in various locations inside and outside of the building. With the increased use of plastic water piping and with raceways, grounding electrodes, and conductors often concealed, locating proper bonding and grounding connections for these other systems becomes problematic. In some cases, communications and CATV installers drive a separate ground rod, possibly creating a potential difference between the different grounding systems. To ensure that these systems are properly bonded, the code now requires installation of an *intersystem bonding termination*, a device that provides a means for connecting grounding and bonding conductors of communications systems near the building service equipment. For separate buildings supplied by a feeder or branch circuit, the device is installed near the disconnecting means. The bonding termination must be accessible and must have provisions for connection of at least three bonding conductors. A set of listed terminals mounted directly to the meter cabinet satisfies these requirements and may be the most common installation. Other approved methods for intersystem bonding include the installation of a bonding bar near the service enclosure, meter cabinet, or service raceway. The bonding bar may also be attached to the grounding electrode conductor.

E3705.7
Location of Overcurrent Devices

CHANGE TYPE: Modification

CHANGE SUMMARY: The code now specifically prohibits overcurrent devices from being located over steps.

2009 CODE: E3705.7 Location of Overcurrent Devices in or on Premises.
Overcurrent devices shall:
1. Be readily accessible.
2. Not be located where they will be exposed to physical damage.
3. Not be located where they will be in the vicinity of easily ignitible material such as in clothes closets.
4. Not be located in bathrooms.
5. <u>Not be located over steps of a stairway.</u>
~~5.~~ 6. Be installed so that the center of the grip of the operating handle of the switch or circuit breaker, when in its highest position, is not more than 6 feet 7 inches (2007 mm) above the floor or working platform.

Exceptions: (No change to text.)

CHANGE SIGNIFICANCE: This change adds a prohibited location for overcurrent devices. Previously, there was no specific language regarding panelboards with circuit breakers or fuses located above steps. Clearly such an installation creates difficulties for servicing, and the height limitations are difficult to measure. Such uncertainties might lead to inconsistent application of the code provisions. Because of the unsafe working conditions that these installations create, overcurrent protection devices are now specifically prohibited above steps. This does not preclude the installation of a panelboard at a landing of a stairway, provided the working clearance requirements are satisfied.

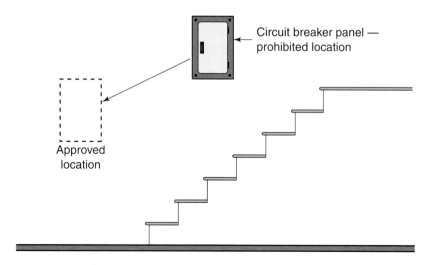

Overcurrent protection devices not permitted over steps of a stairway

CHANGE TYPE: Modification

CHANGE SUMMARY: Receptacle outlets that are controlled by a wall switch are no longer counted as contributing to the required receptacle distribution in the specified rooms or areas.

2009 CODE: E3901.1 General. Outlets for receptacles rated at 125 volts, 15 and 20 amperes, shall be provided in accordance with Sections E3901.2 through E3901.11. The ~~R~~ receptacles ~~outlets~~ required by this section shall be in addition to any receptacle that is:

1. part of a luminaire or appliance
2. ~~that is~~ located within cabinets or cupboards
3. <u>controlled by a wall switch in accordance with Section E3903.2, Exception 1</u>
4. ~~that is~~ located over 5.5 feet (1676 mm) above the floor

Permanently installed electric baseboard heaters equipped with factory-installed receptacle outlets, or outlets provided as a separate assembly by the baseboard manufacturer shall be permitted as the required outlet or outlets for the wall space utilized by such permanently installed heaters. Such receptacle outlets shall not be connected to the heater circuits.

CHANGE SIGNIFICANCE: Wall switch controlled receptacle outlets are permitted for satisfying the minimum lighting requirements for habitable rooms except kitchens as stated in Section E3903. Previously, the code contained no language to exclude these switched receptacle outlets when determining compliance with the spacing and distribution provisions of Section E3901. With the additional language in Section E3901.1, the code clearly does not recognize switched receptacles as required general purpose convenience receptacles. This change may not have a significant impact. If only one of the two receptacles in a duplex outlet is switched, the remaining receptacle with continuous power may be used in satisfying the maximum spacing requirements. On the other hand, if both receptacles are switched, the duplex outlet is not counted.

E3901

Required Receptacle Outlets

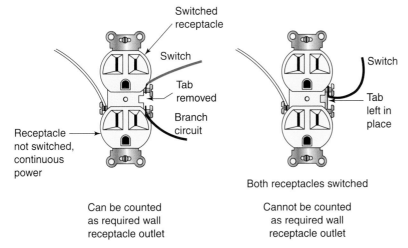

Can be counted as required wall receptacle outlet

Both receptacles switched

Cannot be counted as required wall receptacle outlet

Required receptacle outlets

E3901.7
Outdoor Outlets

CHANGE TYPE: Modification

CHANGE SUMMARY: A receptacle outlet is now required for each balcony, deck, or porch 20 square feet or greater in area.

2009 CODE: E3901.7 Outdoor Outlets.   At least one receptacle outlet that is accessible while standing at grade level and located not more than 6 feet, 6 inches (1981 mm) above grade, shall be installed outdoors at the front and back of each dwelling unit having direct access to grade. Balconies, decks, and porches that are accessible from inside of the dwelling unit and that have a usable area of 20 square feet (1.86 m^2) or greater shall have at least one receptacle outlet installed within the perimeter of the balcony, deck, or porch. The receptacle shall be located not more than 6 feet, 6 inches (1981mm) above the balcony, deck, or porch surface.

CHANGE SIGNIFICANCE: In addition to the required outdoor receptacle outlets at the front and back of the dwelling, the code now requires at least one receptacle outlet at each balcony, deck, or porch that is accessible from inside the dwelling and is 20 square feet or greater in area. The new provision applies whether or not the deck, porch, or balcony has access to grade. The maximum height requirement of 6 feet 6 inches above the walking surface matches that of the other outdoor outlets. This change is in response to concerns about use of extension cords for these outdoor areas. Balconies or decks less than 20 square feet are typically for architectural or decorative purposes, and requiring a receptacle outlet was considered unnecessary.

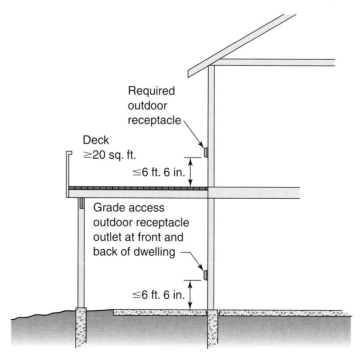

Required outdoor receptacle

Deck ≥20 sq. ft.

≤6 ft. 6 in.

Grade access outdoor receptacle outlet at front and back of dwelling

≤6 ft. 6 in.

Outdoor receptacle outlets

CHANGE TYPE: Modification

CHANGE SUMMARY: Ground-fault circuit-interrupter protection is now required for all 125-volt, single-phase, 15- and 20-ampere receptacles installed in garages and unfinished basement areas except those for fire or burglar alarm systems.

2009 CODE: E3902.2 Garage and Accessory Building Receptacles.   All 125-volt, single-phase, 15- or 20-ampere receptacles installed in garages and grade-level portions of unfinished accessory buildings used for storage or work areas shall have ground-fault circuit-interrupter protection for personnel.

~~Exceptions:~~
~~1. Receptacles that are not readily accessible.~~

~~2. A single receptacle or a duplex receptacle for two appliances located within dedicated space for each appliance that in normal use is not easily moved from one place to another, and that is cord- and plug-connected.~~

E3902.2 and E3902.5 continues

E3902.2 and E3902.5

Ground-Fault Circuit-Interrupter (GFCI) Protection

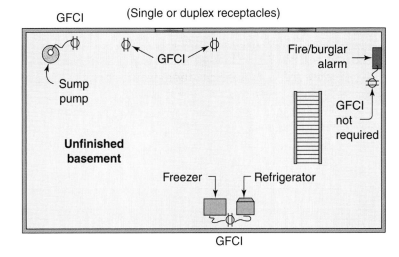

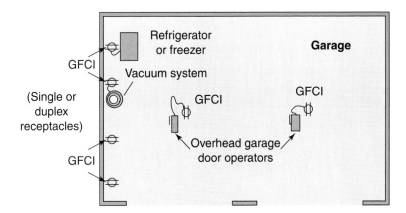

GFCI-protected receptacle outlets

E3902.2 and E3902.5 continued

E3902.5 Unfinished Basement Receptacles. All 125-volt, single-phase, 15- and 20-ampere receptacles installed in unfinished basements shall have ground-fault circuit-interrupter protection for personnel. For purposes of this section, unfinished basements are defined as portions or areas of the basement not intended as habitable rooms and limited to storage areas, work areas, and the like.

Exceptions:

1. ~~Receptacles that are not readily accessible.~~

2. ~~A single receptacle or duplex receptacle for two appliances located within dedicated space for each appliance that in normal use is not easily moved from one place to another, and that is cord- and plug-connected.~~

3. A receptacle supplying only a permanently installed fire alarm or burglar alarm system.

CHANGE SIGNIFICANCE: Previously, receptacles installed in garages and unfinished basement areas required ground-fault circuit-interrupter (GFCI) protection, but the code offered exceptions for receptacles that were not accessible and those located in spaces dedicated for appliances. Other than receptacles serving an alarm system, the exceptions from GFCI protection in unfinished basements and garages have been removed. Single receptacles serving sump pumps, refrigerators, freezers, or similar appliances in these locations are no longer exempt and require GFCI protection. Appliances are manufactured to be compatible with GFCI devices, and nuisance tripping is no longer the concern it once was.

Similarly, receptacle outlets on garage ceilings for overhead door operators and other locations that previously have been considered not readily accessible now require GFCI protection. The term *readily accessible* was considered ambiguous and subjective. In addition, appliances are often in a location that is readily accessible even if plugged into an outlet that is not. Changes to this section expand the GFCI requirements and afford an increased level of safety for dwelling occupants.

E3902.11
Arc-Fault Protection

CHANGE TYPE: Modification

CHANGE SUMMARY: Arc-fault protection for branch circuits has been expanded to include all habitable spaces (except kitchens), hallways, closets, and similar areas. Only a combination type arc-fault circuit interrupter is permitted and it must protect the entire branch circuit.

2009 CODE: E3902.11 Arc-Fault Circuit-Interrupter Protection of Bedroom Outlets.   All branch circuits that supply 120-volt, single-phase, 15- and 20-ampere outlets installed in <u>family rooms, dining rooms, living rooms, parlors, libraries, dens,</u> bedrooms, <u>sunrooms, recreation rooms, closets, hallways, or similar rooms or areas</u> shall be protected by a combination type ~~or branch/feeder type~~ arc-fault circuit interrupter installed to provide protection of the entire branch circuit. ~~Effective January 1, 2008, such arc-fault circuit interrupter devices shall be combination type.~~

> **Exception:** ~~The location of the arc-fault circuit interrupter shall be permitted to be at other than the origination of the branch circuit provided that:~~
> 1. ~~The arc-fault circuit interrupter is installed within 6 feet (1.8 m) of the branch circuit overcurrent device as measured along the branch circuit conductors and~~
> 2. ~~The circuit conductors between the branch circuit overcurrent device and the arc-fault circuit interrupter are installed in a metal raceway or a cable with a metallic sheath.~~

CHANGE SIGNIFICANCE: Arc-fault circuit-interrupter (AFCI) protection is now required for branch circuits serving outlets in most living areas of a dwelling. Because of the different functions and electrical load requirements, kitchens, bathrooms, unfinished basements, ga-

E3902.11 continues

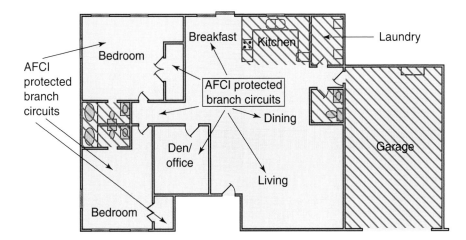

ARC-fault protection

E3902.11 continued rages, and outdoor outlets do not require AFCI protection. Previously, only circuits serving outlets in bedrooms required protection. The expanded coverage is the result of increasing confidence in the evolving AFCI technology and proven effectiveness of these devices in detecting arcing characteristics and preventing fires. The code permits only combination type AFCI devices, which are tested and listed for both branch/feeder and outlet circuit protection. Prior to January 1, 2008, the code permitted the installation of either a combination type or a branch/feeder type of AFCI device. Combination type devices provide an increased level of protection for cord sets that are plugged into receptacles on an AFCI-protected circuit.

CHANGE TYPE: Addition

CHANGE SUMMARY: The code now requires listed tamper-resistant receptacles for all 125-volt 15- and 20-ampere receptacles installed in dwelling units, on the outside of dwelling units, and in attached and detached garages.

2009 CODE: E4002.14 Tamper-Resistant Receptacles. In areas specified in Section E3901.1, 125-volt, 15- and 20-ampere receptacles shall be listed tamper-resistant receptacles.

CHANGE SIGNIFICANCE: Tamper-resistant receptacles are designed to prevent the insertion of any small object, such as a paper clip, into one side of the receptacle. Both blades of an attachment plug must be inserted simultaneously to open the protective shield and allow connection to electricity. The code now requires that all 125-volt 15- and 20-ampere receptacles installed in areas mentioned in Section E3901.1 be listed as tamper resistant. This new requirement applies to all areas of dwellings and garages, including the required outdoor outlets. This added safeguard in the electrical provisions intends to reduce the number of electrical shock injuries to children.

E4002.14

Tamper-Resistant Receptacles

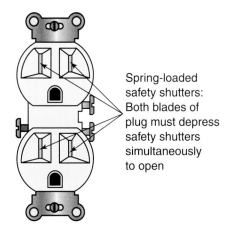

Spring-loaded safety shutters: Both blades of plug must depress safety shutters simultaneously to open

Tamper resistant receptacles

E4003.12

Luminaires in Clothes Closets

CHANGE TYPE: Modification

CHANGE SUMMARY: Recessed or surface-mounted LED luminaires are now permitted in locations approved for incandescent luminaires. Surface-mounted LED and fluorescent luminaires may be installed within the defined storage area when the luminaire is specifically identified as suitable for such use.

2009 CODE: E4003.12 Luminaires in Clothes Closets. For the purposes of this section, storage space shall be defined as a volume bounded by the sides and back closet walls and planes extending from the closet floor vertically to a height of 6 feet (1829 mm) or the highest clothes-hanging rod and parallel to the walls at a horizontal distance of 24 inches (610 mm) from the sides and back of the closet walls respectively, and continuing vertically to the closet ceiling parallel to the walls at a horizontal distance of 12 inches (305 mm) or the width of the shelf, whichever is greater. For a closet that permits access to both sides of a hanging rod, the storage space shall include the volume below the highest rod extending 12 inches (305 mm) on either side of the rod on a plane horizontal to the floor extending the entire length of the rod (see Figure E4003.12). The types of luminaires installed in clothes closets shall be limited to surface-mounted or recessed incandescent luminaires with completely enclosed lamps, ~~and~~ surface-mounted or recessed fluorescent luminaires, <u>and surface-mounted fluorescent or LED luminaires identified as suitable for installation within the storage area</u>. Incandescent luminaires with open or partially enclosed lamps and pendant luminaires or lamp-holders shall be prohibited. ~~Luminaire installations shall be in accordance with one or more of the following~~ <u>The minimum clearance between luminaires installed in clothes closets and the nearest point of a storage area shall be as follows:</u>

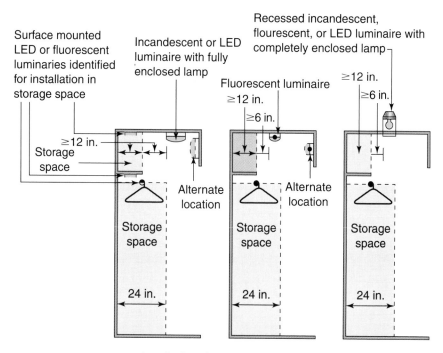

Luminaires in clothes closets

1. Surface-mounted incandescent <u>or LED luminaires with a completely enclosed light source</u> shall be installed on the wall above the door or on the ceiling, provided <u>that</u> there is a minimum clearance of 12 inches (305 mm) ~~between the fixture and the nearest point of a storage space~~.

2. Surface-mounted fluorescent luminaires shall be installed on the wall above the door or on the ceiling, provided <u>that</u> there is a minimum clearance of 6 inches (152 mm) ~~between the fixture and the nearest point of a storage space~~.

3. Recessed incandescent <u>or LED</u> luminaires with a completely enclosed <u>light source</u> ~~lamp~~ shall be installed in the wall or the ceiling provided there is a minimum clearance of 6 inches (152 mm) ~~between the fixture and the nearest point of a storage space~~.

4. Recessed fluorescent luminaires shall be installed in the wall or on the ceiling provided there is a minimum clearance of 6 inches (152 mm) ~~between the fixture and the nearest point of a storage space~~.

5. <u>Surface-mounted fluorescent or LED luminaires shall be permitted to be installed within the storage space where identified for this use.</u>

CHANGE SIGNIFICANCE: This change recognizes the new technology of LED lighting, which is designed for low heat output. Clearances for LED luminaires in clothes closets generally match the permitted locations for incandescent luminaires—6 inches from the storage space for recessed fixtures and 12 inches from the storage space for surface-mounted fixtures. However, when specifically identified as suitable for installation within the closet storage space, both surface-mounted LED luminaires and surface-mounted fluorescent luminaires may be installed anywhere in a clothes closet in accordance with the manufacturer's installation instructions. These fixtures are designed with sufficiently low heat output that they do not pose an ignition hazard for combustibles in storage areas.

E4203.3

Disconnecting Means for Pools, Spas, and Hot Tubs

CHANGE TYPE: Modification

CHANGE SUMMARY: The required disconnecting means for pools, spas, and hot tubs must simultaneously disconnect all ungrounded conductors and be located not less than 5 feet from the water's edge.

2009 CODE: E4203.3 Disconnecting Means. One or more means to <u>simultaneously</u> disconnect all ungrounded conductors for all utilization equipment, other than lighting, shall be provided. Each of such means shall be readily accessible and within sight from the equipment it serves <u>and shall be located at least 5 feet (1524 mm) horizontally from the inside walls of a pool, spa, or hot tub unless separated from the open water by a permanently installed barrier that provides a 5 foot (1524 mm) or greater reach path. This horizontal distance shall be measured from the water's edge along the shortest path required to reach the disconnect.</u>

CHANGE SIGNIFICANCE: Section E4203.3 requires a disconnecting means for all utilization equipment, such as pumps or heaters, associated with pools, spas, or hot tubs. The disconnect allows for safe servicing, maintenance, and replacement of the utilization equipment and must be located within sight of such equipment. The code now specifically requires a horizontal separation of not less than 5 feet between the disconnecting means and the inside wall of the pool, spa, or hot tub. The 5-foot distance is consistent with other clearances in the electrical provisions and matches the separation requirements for switching devices in Section E4203.2.

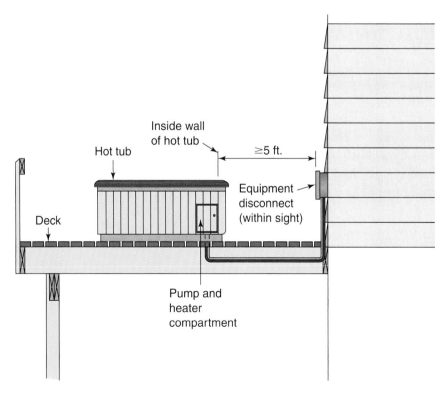

Disconnecting means for pools, spas, and hot tubs

CHANGE TYPE: Clarification

CHANGE SUMMARY: Revisions to Section E4206.5.1 clarify that wet-niche luminaires must be installed such that all maintenance can be performed on the pool deck.

2009 CODE: E4206.5.1 Servicing.   All <u>wet-niche</u> luminaires shall be removable from the water for <u>inspection,</u> relamping or <u>other</u> ~~normal~~ maintenance. <u>The forming shell location and length of cord in the forming shell shall permit personnel to place the removed luminaire on the deck or other dry location for such maintenance. The luminaire maintenance location shall be accessible without entering or going into the pool water.</u> ~~Luminaires shall be installed in such a manner that personnel can reach the luminaire for relamping, maintenance, or inspection while on the deck or equivalently dry location.~~

CHANGE SIGNIFICANCE: Wet-niche luminaires must be installed in a location that provides access and a means for inspection, maintenance, or changing lamps. Revisions to this section clarify that access and removal of the luminaire must be accomplished from a position on the deck, grade, or other dry surface adjoining the pool without entering the pool. In addition, the fixture cord must be of sufficient length to place the luminaire on an appropriate work surface, such as the deck, for maintenance. The intent of the change is to ensure a safe means for servicing of this pool equipment.

E4206.5.1
Servicing of Wet-Niche Luminaires

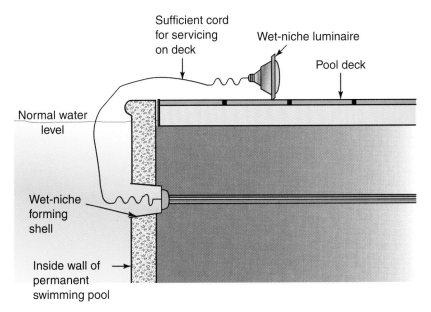

Servicing of wet-niche luminaires

E4209.1

Hydromassage Bathtubs

CHANGE TYPE: Modification

CHANGE SUMMARY: Hydromassage bathtubs are now required to be on an individual branch circuit. The required ground-fault circuit-interrupter (GFCI) device must be readily accessible.

2009 CODE: E4209.1 Ground-Fault Circuit Interrupters. Hydromassage bathtubs and their associated electrical components shall be ~~protected in accordance with Section E4108~~ <u>supplied by an individual branch circuit(s) and protected by a readily accessible ground-fault circuit interrupter</u>. All 125-volt, single-phase receptacles not exceeding 30 amperes and located within ~~5~~ <u>6</u> feet (~~1524~~ <u>1829</u> mm) measured horizontally of the inside walls of a hydromassage tub shall be protected by a ground-fault circuit interrupter(s).

CHANGE SIGNIFICANCE: A hydromassage bathtub, commonly referred to as a whirlpool tub, is by definition a permanently installed bathtub equipped with a recirculating piping system, pump, and associated equipment. It is designed so it can accept, circulate, and discharge water upon each use. Previously, the code required GFCI protection for the power supply to the hydromassage bathtub but did not stipulate the location of the GFCI device. In many cases, a GFCI receptacle outlet for the pump motor was installed behind the access panel and in the vicinity of the pump. The code now specifically requires the GFCI device to be in a readily accessible location. Readily accessible means that access is available without removing a panel

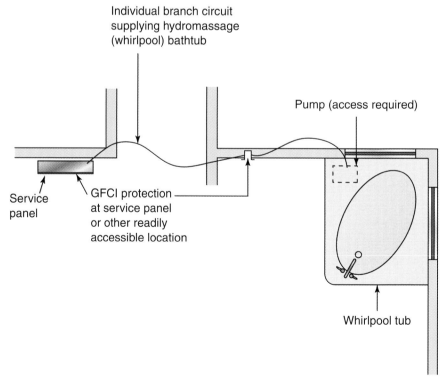

Individual branch circuit supplying hydromassage (whirlpool) bathtub

Pump (access required)

Service panel

GFCI protection at service panel or other readily accessible location

Whirlpool tub

Hydromassage bathtubs

or obstruction. The intent is to ensure that the GFCI protection and its location are apparent to the homeowner and others.

At least one individual branch circuit is now required to serve a hydromassage bathtub. This is consistent with most manufacturers' installation instructions and intends to prevent the practice of supplying pump motors or heaters from other branch circuits serving receptacle outlets located in the bathroom or in other areas of the dwelling. By definition, an individual branch circuit supplies only one piece of utilization equipment. The other change to this section increases the minimum separation distance to 6 feet between the inside walls of the tub and a receptacle outlet that is not GFCI protected. For 125-volt 30-amp receptacles less than 6 feet from the tub, GFCI protection is required. This change is not likely to have a significant impact on installations, because all 125-volt 15- and 20-amp receptacles in bathrooms require GFCI protection.

PART 9

Swimming Pools, Spas, and Hot Tubs

Appendix G

Provisions in the appendices of the IRC do not apply unless specifically referenced in the adopting ordinance of the jurisdiction. When adopted, Appendix G sets requirements for swimming pools, spas and hot tubs on the premises of one- and two-family dwellings. Sections AG102 through AG104 provide definitions specific to Appendix G and reference approved standards for the design and construction of pools, spas and hot tubs. The balance of this appendix chapter sets barrier requirements to prevent small children from accessing pool areas unattended and entrapment protection requirements to further improve pool safety. ■

AG105.2
Outdoor Swimming Pool Barrier

AG106
Entrapment Protection for Swimming Pool and Spa Suction Outlets

AG105.2

Outdoor Swimming Pool Barrier

CHANGE TYPE: Clarification

CHANGE SUMMARY: For doors that provide access to an outdoor swimming pool, the prescriptive requirements for door alarm activation, deactivation, and reset have been deleted in favor of matching requirements in UL 2017.

2009 CODE: AG105.2 Outdoor Swimming Pool. An outdoor swimming pool, including an in-ground, above-ground, or on-ground pool, hot tub or spa shall be surrounded by a barrier which shall comply with the following:

Items 1. (through) 8.   (No change to text.)

  9. Where a wall of a dwelling serves as part of the barrier, one of the following conditions shall be met:

  9.1. The pool shall be equipped with a powered safety cover in compliance with ASTM F 1346; or

  9.2. Doors with direct access to the pool through that wall shall be equipped with an alarm which produces an audible warning when the door and/or its screen, if present, are opened. The alarm shall be listed <u>and labeled </u>in ac-

AG105.2 continues

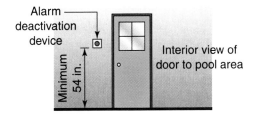

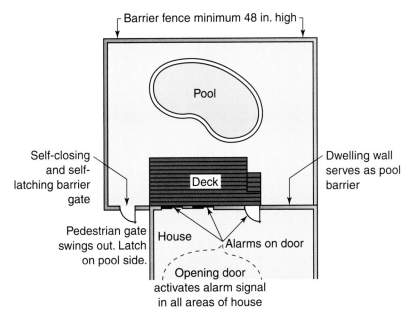

Pool barrier and door alarm requirements

AG105.2 continued

cordance with UL 2017. ~~The audible alarm shall activate within 7 seconds and sound continuously for a minimum of 30 seconds after the door and/or its screen, if present, are opened and be capable of being heard throughout the house during normal household activities. The alarm shall automatically reset under all conditions. The alarm system shall be equipped with a manual means, such as touch pad or switch, to temporarily deactivate the alarm for a single opening. Deactivation shall last for not more than 15 seconds.~~ The deactivation switch(es) shall be located at least 54 inches (1372 mm) above the threshold of the door; or

9.3. Other means of protection, such as self-closing doors with self-latching devices, which are approved by the governing body, shall be acceptable so long as the degree of protection afforded is not less than the protection afforded by Item 9.1 or 9.2 described above.

10. (No change to text.)

CHANGE SIGNIFICANCE: An outdoor swimming pool is typically accessed from the dwelling unit through a patio door or other exterior door. One of three options for satisfying the barrier requirements related to such access is to install an alarm on the door. The 2006 IRC prescribed the alarm operation criteria, including means for activation, temporary deactivation, and automatic reset. The 2009 IRC deletes the alarm operation specifications because such information is contained in the referenced standard UL 2017 *Standard for Safety General-Purpose Signaling Devices and Systems.* UL 2017 covers residential *water hazard entrance alarms*, devices or systems intended to be installed on gates, doors, or access barriers surrounding residential swimming pools, spas, or hot tubs for the purpose of sounding an audible alarm should a young child gain entry into these areas. UL 2017 includes the requirements identified in the code as well as an operation test, an audibility test, and a static discharge test. The 2009 IRC clarifies that the alarm must be both *listed* and *labeled* as conforming to UL 2017. The code maintains the minimum 54-inch height requirement for the deactivation switch to keep it out of reach of small children.

AG106

Entrapment Protection for Swimming Pool and Spa Suction Outlets

CHANGE TYPE: Modification

CHANGE SUMMARY: The provisions to avoid entrapment hazards at suction outlets have been deleted in favor of the referenced standard ANSI/APSP-7 *Standard for Suction Entrapment Avoidance in Swimming Pools, Wading Pools, Spas, Hot Tubs, and Catch Basins.*

2009 CODE: **AG106.1 General.** Suction outlets shall be designed <u>and installed in accordance with ANSI/APSP-7.</u>

~~**AG106.2 Suction Fittings.** Pool and spa suction outlets shall have a cover that conforms to ANSI/ASME A112.19.8M, or an 18-inch×23 inch (457mmby 584 mm) drain grate or larger, or an approved channel drain system.~~

~~**Exception:** Surface skimmers~~

~~**AG106.3 Atmospheric Vacuum Relief System Required.** Pool and spa single- or multiple-outlet circulation systems shall be equipped with atmospheric vacuum relief should grate covers located therein become missing or broken. This vacuum relief system shall include at least one approved or engineered method of the type specified herein, as follows:~~

~~1. Safety vacuum release system conforming to ASME A112.19.17; or~~

~~2. An approved gravity drainage system.~~

~~**AG106.4 Dual Drain Separation.** Single or multiple pump circulation systems shall be provided with a minimum of two suction outlets of the approved type. A minimum horizontal or vertical distance of 3 feet (914 mm) shall separate the outlets. These suction outlets shall be piped so that water is drawn through them simultaneously through a vacuum-relief-protected line to the pump or pumps.~~

~~**AG106.5 Pool Cleaner Fittings.** Where provided, vacuum or pressure cleaner fitting(s) shall be located in an accessible position(s) at least 6 inches (152 mm) and not more than 12 inches (305 mm) below the minimum operational water level or as an attachment to the skimmer(s).~~

Section AG108 Standards

<u>ANSI/APSP-7-06 Standard for Suction Entrapment Avoidance in Swimming Pools, Wading Pools, Spas, Hot Tubs, and Catch Basins</u>

ASME/ANSI A112.19.~~8M-1987(R1996)~~ <u>8-2007</u> Suction Fittings for Use in Swimming Pools, Wading Pools, Spas, Hot Tubs, and Whirlpool Bathing Appliances

CHANGE SIGNIFICANCE: This change adds a new standard, ANSI/APSP-7 *Standard for Suction Entrapment Avoidance in Swimming Pools, Wading Pools, Spas, Hot Tubs, and Catch Basins*, into IRC

AG106 continues

AG106 continued   Appendix Section AG 106 to provide a comprehensive approach to entrapment hazards and improve pool safety. It also deletes Sections AG106.2 through AG106.6 because all of these requirements have been incorporated into ANSI/APSP-7. The standard was developed by the Association of Pool and Spa Professionals (APSP) based on the latest information and technology to address all forms of entrapment hazards and their underlying causes, and intends to be consistent with similar federal legislation. This standard provides that all swimming pools and spas be equipped with proper anti-entrapment drain covers and circulation and drainage systems. Appendix G is in effect only if specifically referenced in the adopting legislation of the jurisdiction.

Index

INTRODUCING SAVE™

Sustainable **A**ttributes **V**erification and **E**valuation™—New from ICC-ES®

The new SAVE™ program from ICC-ES® provides the most trusted third-party verification available today for sustainable construction products. Under this program, ICC-ES evaluates and confirms product's sustainable attributes. The SAVE™ program may also assist in identifying products that help qualify for points under major green rating systems such as US Green Building Council's LEED, Green Building Initiative's Green Globes or ICC/NAHB's proposed National Green Building Standard (NGBS). When it comes to making sure that products possess the sustainable attributes claimed, you can trust ICC-ES SAVE.

FOR MORE INFORMATION ABOUT SAVE: 1-800-423-6587 | www.icc-es.org/save

ICC EVALUATION SERVICE

8-61804-66

Approve plumbing products with a name you have come to trust.

ICC-ES PMG Listing Program

When it comes to approving plumbing, mechanical, or fuel gas (PMG) products, ask manufacturers for their ICC-ES PMG Listing. Our listing ensures code officials that a thorough evaluation has been done and that a product meets the requirements in both the codes and the standards. ICC-ES is the name code officials prefer when it comes to approving products.

FOR DETAILS! 1-800-423-6587, x5478 | www.icc-es.org/pmg